CATALYST

A framework for success

Carol Chapman
Moira Sheehan

Heinemann
Inspiring generations

Contents

T indicates Think about spread

iii

Introduction

Welcome to Catalyst

This is the first of three books designed to help you learn all the science ideas you need during Key Stage 3. We hope you'll enjoy the books as well as learning a lot from them.

This book has twelve units which each cover a different topic. The units have two types of pages:

Learn about:

Most of the double-page spreads in a unit introduce and explain new ideas about the topic. They start with a list of these so that you can see what you are going to learn about.

Think about:

Each unit has a double-page spread called Think about. You will work in pairs or small groups and discuss your answers to the questions. These pages will help you understand how scientists work and how ideas about science develop.

On the pages there are these symbols:

a Quick questions scattered through the pages help you check your knowledge and understanding of the ideas as you go along, for example,

> **a** **Use the particle model to explain why the liquid will not squash .**

Questions

The questions at the end of the spread help you check you understand all the important ideas.

For your notes:

These list the important ideas from the spread to help you learn, write notes and revise.

Do you remember?

These remind you of what you already know about the topic.

Did you know?

These tell you interesting or unusual things, such as the history of some science inventions and ideas.

At the back of the book:

Glossary

All the important scientific words in the text appear in bold type. They are listed with their meanings in the Glossary at the back of the book. Look there to remind yourself what they mean.

Index

There is an Index at the very back of the book, where you can find out which pages cover a particular topic.

Activities to help or check your learning:

Your teacher may give you these activities from the teacher's materials which go with the course:

Transition quiz and worksheets

Before you start a new unit your teacher may give you a quiz or a worksheet. This short exercise will help you remember what you already know about a topic.

Homework

At the end of a lesson the teacher may give you one of the homework sheets that go with the lesson. This will help you to review and revise what you learnt in the lesson.

Unit map

You can use this to think about what you already know about a topic. You can also use it to revise a topic before a test or exam.

Pupil check list

This is a check list of what you should have learnt to help you with your revision.

Starters

When you start a lesson this is a short activity to introduce what you are going to learn about.

Test yourself

You can use this quiz at the end of each unit to see what you are good at and what you might need to revise.

Activity

There are different types of activities, including investigations, that your teacher can give you to help with the topics in each spread in the pupil book.

End of unit test Tier 2–5

This helps you and your teacher check what you learnt during the unit, and measures your progress and success.

Plenaries

At the end of a lesson your teacher may give you a short activity to summarise what you have learnt.

Heinemann Educational Publishers
Halley Court, Jordan Hill, Oxford OX2 8EJ
Part of Harcourt Education

Heinemann is the registered trademark of
Harcourt Education Limited

© Carol Chapman, Moira Sheehan 2003

First published 2003

07 06 05 04 03
10 9 8 7 6 5 4 3 2 1

British Library Cataloguing in Publication Data is available
from the British Library on request.

ISBN 0 435 76011 4

Edited by Ruth Holmes and Sarah Ware
Designed by Ken Vail Graphic Design
Typeset by Ken Vail Graphic Design

Original illustrations © Harcourt Education Limited 2003

Illustrated by Graham-Cameron Illustration (Tim Archbold, Darin Mount and Sarah
Wimperis), Nick Hawken, Illustration Ltd (David Ashby), B. L. Kearley (Sheila Galbreath,
Jeremy Gower and Pat Tourett), Linda Rogers Associates (Lorna Barnard, Dave
Burroughs, Keith Howard, Gary Rees and Branwen Thomas), David Lock, Joseph McEwan,
John Plumb, Simon Girling & Associates (Mike Lacey), Sylvie Poggio Artists Agency (Nigel
Kitching, Rhiannon Powell, Lisa Smith and Sean Victory)

Printed in the UK by Bath Press Ltd

Picture research by Jennifer Johnson

Acknowledgements
The authors and publishers would like to thank the following for permission to use
copyright material: **bar chart p92**, Addison-Wesley, *Body Maintenance*; **graph p47**,
The Canadian Journal of Rural Medicine Vol. 3, p12–19, with permission from the Society
of Rural Physicians of Canada; **map p8**. This product includes mapping data licensed from
Ordnance Survey® with permission of the Controller of her Majesty's Stationary Office, ©
Crown copyright. All rights reserved. License no. 100000230.

The publishers have made every effort to trace the copyright holders, but if they have
inadvertently overlooked any, they will be pleased to make the necessary
arrangements at the first opportunity.

For photograph acknowledgements, please see page vii.

Tel: 01865 888058 www.heinemann.co.uk

The author and publishers would like to thank the following for permission to use photographs:

T = top   **B** = bottom   **L** = left   **R** = right   **M** = middle

SPL = Science Photo Library

Cover: Getty Images.

Page 2, **L**: Biophoto Associates; 2, **R**: SPL/M.I. Walker; 3: SPL/Eye of Science; 6, x3: Biophoto Associates; 7, x3: Biophoto Associates; 10: SPL/Andrew Syred; 12: Bruce Coleman; 14, **T**: SPL/Science Pictures; 14, **B**: SPL/Andy Walker; 15, **T**: SPL/D. Phillips; 15, **B**: SPL/Don Fawcett; 16, **T**: SPL/Dr. Yorgos Nikas; 16, **M**: SPL/Dr. Yorgos Nikas; 16, **B1**: SPL/Profs PM Motta & S Makeabe; 16, **B2**: SPL/Petit Format, Nestle; 16, **B3**: SPL/Alex Bartel; 16, **B**4: S&R Greenhill; 19, **L**: S&R Greenhill; 19, **R**: S&R Greenhill; 20 S&R Greenhill; 22: Bruce Coleman; 22 inset: NHPA/A.N.T.; 23, **T**: NHPA/A Warburton & S. Toon; 23, **B**: Bruce Coleman; 24, **TL**: SPL/Bernard Edmaier; 24, **TM**: NHPA/James Carmichael Jr; 24, **TR** Hilary Fletcher; 24, **B**: SPL/Tony Craddock; 25, **T**: SPL/Tom McHugh; 25, **BL**: FLPA/Philip Perry; 25, **BR**: FLPA/Jurgen & Christine Sohns; 26: FLPA/Images of Nature, John Hawkins; 27, **T**: Heather Angel; 27, **B**: FLPA/D P Wilson; 28, **TL**: FLPA/David Hosking; 28, **TR**: FLPA/M J Thomas; 28, **BL**: Holt Studios/Nigel Cattlin; 28, **BR**: Holt Studios/Nigel Cattlin; 29, **T**: OSF/Tim Shepherd; 29, **M**: Corbis/Steve Kaufman; 29, **B**: SPL/Brock May; 30, **L**: Holt Studios/Nigel Cattlin; 30, **R**: Bruce Coleman/ Sarah Cook; 31, **T**: NHPA/Dave Watts; 31, **B**: Bruce Coleman/Paul van Gaalen; 35, **T1**: NHPA/Roy Waller; 35, **T**2: Heather Angel; 35, **T**3: Heather Angel; 35, **T**4: NHPA/Stephen Dalton; 35, **B1**: Heather Angel; 35, **B2**: NHPA/Trevor McDonald; 35, **B3**: FLPA/Mike J Thomas; 35, **B**4: NHPA/Martin Harvey; 36, **L**: NHPA/Darek Karp; 36, **R**: OSF/Johnny Johnson; 37: Tony Stone; 40, **LT**: Still Pictures/Ron Gilling; 40, **LM**: Still Pictures/Francois Gilson; 40, **LB**: OSF/Harry Taylor; 40, **M1**: Hans Reinhard; 40, **M2**: OSF/John Downer; 40, **M3**: OSF/Scott Camazine/CDC; 40, **M4**: SPL/Dr Linda Stannard; 40, **RT**: OSF/David M Dennis; 40, **RB**: Bruce Coleman; 41: MEPL; 42, **T1**: Photodisc; 42, **T2**: Photodisc; 42, **T3**: OSF/Maurice Tibbles; 42, **T4**: Photodisc; 42, **T5**: Bruce Coleman; 42, **B**: FLPA/Images of Nature/ John Hawkins; 43, **T**: OSF/Max Gibbs; 43, **M**: OSF/Souricat; 43, **B**: Bruce Coleman; 44, **TL**: OSF/Fredrik Ehren Strom; 44, **BL**: Bruce Coleman; 44, **TM**: OSF/Rudie Kuiter; 44, **BM**: OSF/London Scientific Film; 44, **TR**: OSF/David Fox; 44, **BR**: FLPA/Images of Nature/Gerard Laci; 45, **TL**: OSF/Frank Schneidermever; 45, **TR**: Bruce Coleman; 45, **ML**: Bruce Coleman/Animal Ark; 45, **MR**: OSF/Mills Tandy; 45, **B**: SPL/Sinclair Stammers; 48, **T**: Andrew Lambert; 48, **B**: Andrew Lambert; 49: Andrew Lambert; 50, x50: Andrew Lambert; 52, **T**: Andrew Lambert; 52, **B**: Andrew Lambert; 53: Garden and Wildlife Matters; 54: Holt Studios/Primrose Peacock; 56: Robert Harding Picture Library; 57, **T**: Pete Morris; 57, **M**: Robert Harding Picture Library; 57, **B**: Peter Gould; 58, **T**: Andrew Lambert; 58, **B**: Andrew Lambert; 59: Roger Scruton; 60, **T**: Andrew Lambert; 60, **B**: Peter Gould; 61, **T**: Anthony Blake Picture Library/Tim Hill; 61, **B**: Source Unknown; 63, **T**: Shout; 63, **B**: Robert Harding/Shout P. Allen; 64, **T**: Tony Stone; 64, **B**: SPL/Jerry Mason; 65, **T**: Gareth Boden; 65, **B**: Andrew Lambert; 66: Robert Harding; 68: Ancient Art & Architecture; 69, **T**: SPL/Sheila Terry; 69, **B**: IBM Corporation, Almaden Research Centre; 70, **T**: Trevor Hill; 70, **B**: Gareth Boden; 71: SPL/Richard Folwell; 73: TRIP/H Rogers; 74, **L**: Andrew Lambert; 74, **M**: Andrew Lambert; 74, **R**: Andrew Lambert; 75, x4: Andrew Lambert; 77, **TL**: Peter Gould; 77, **TR**: Peter Gould; 77, **B**: Barnaby's Picture Library; 78, 1: Roger Scruton; 78, 2: Source Unknown; 78, 3: Andrew Lambert; 78, 4: Andrew Lambert; 79: Roger Scruton; 80: Source Unknown; 81: Peter Gould; 82, x2: Peter Gould; 86, x4: Peter Gould; 87: Gareth Boden; 88, **T**: Empics; 88, **B**: J Allan Cash Ltd; 89: Trevor Hill; 99, **TL**: Tony Stone/Yvette Cardozo; 99, **TR**: Tony Stone/Lorne Resnick; 99, **B**: Holt Studios/Nigel Cattlin; 100, **T**: Environmental Images/David Hoffman; 100, **M**: Environmental Images/David Hoffman; 100, **B**: SPL/Martin Bond; 101, **T**: Tony Stone; 101, **B**: AP/Steve Holland; 106: SPL/Tek Image; 114, **L**: Photodisc; 114, **R**: Action Plus; 115: Peter Morris; 117, **T**: Andrew Lambert; 117, **B**: Andrew Lambert; 124, **TL**: SPL/Jerry Mason; 124, **TM**: SPL/Andrew McClenaghan; 124, **TR**: SPL/Amy Trustram Eve; 124, **B**: SPL/Pekka Parviainen; 125: SPL/Space Telescope Science Institute/NASA; 130, Mercury SPL/NASA, Mehau Kulyk; 130, Venus SPL/NASA; 130, Earth SPL/NASA; 130, Mars SPL/US Geological Society; 130, Jupiter SPL/NASA; 130, Saturn SPL/Space telescope Institute/NASA; 131, Uranus SPL/NASA; 131, Neptune SPL/NASA; 131, Pluto SPL/Space Telescope Institute/NASA.

A1 Organs, cells, tissues

Organs do their job

Animals and plants are made up of **organs**. Each organ has a job.

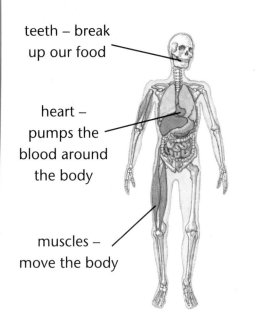

teeth – break up our food

heart – pumps the blood around the body

muscles – move the body

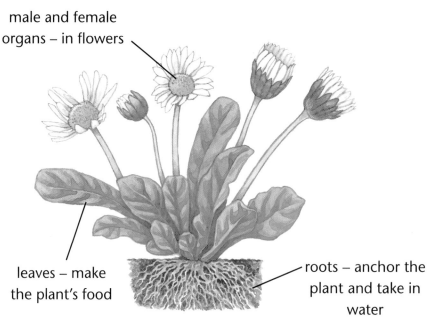

male and female organs – in flowers

leaves – make the plant's food

roots – anchor the plant and take in water

a Which animal organ pumps blood?

b Which plant organ makes food?

Looking closer

Organs in your body are made of smaller parts. You can see this using a **microscope**. A microscope makes small things look much bigger.

Microscopes show that all living things are made up of units called **cells**. Look at the photos below. They show cells from an animal and from a plant.

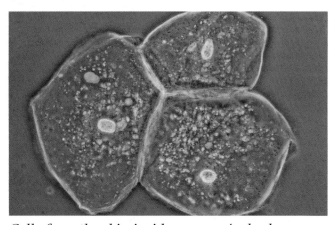

Cells from the skin inside a person's cheek.

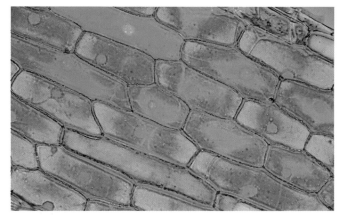

Cells from an onion.

Building an organ from cells

Think about a house. It has a roof, windows and walls. But piles of bricks, panes of glass and roof tiles don't make a house. They have to be arranged so that the bricks make walls, the roof tiles make a roof and the panes of glass make windows.

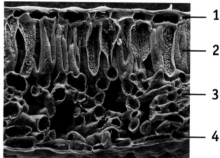

Cells in an organ also have to be arranged. Look at the picture on the right of the cells in a leaf. There are four different layers of cells.

These layers of cells are called **tissues**.

The cells in one tissue are all alike. This is because they have the same job to do. In the picture, all the cells in layer **2** trap sunlight to make food.

C Look at the picture below. Name: (i) the cell (ii) the tissue (iii) the organ (iv) the whole animal.

1 upper epidermis
2 palisade mesophyll layer
3 spongy mesophyll layer
4 lower epidermis

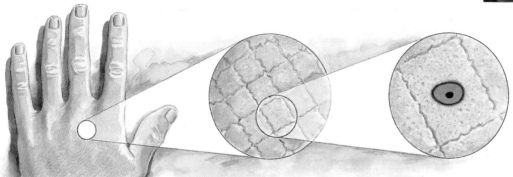

The skin is an **organ**.

The outer layer of skin is a **tissue**.

This is a skin **cell**.

A group of tissues working together makes an organ.

A house is like an organ. The walls and roof are like tissues. The bricks and roof tiles are like cells. They make the walls and roof.

Questions

1 Copy and complete these sentences using the words below.

alike cells organs leaves job roots tissues

Plants have many _____ including stems, _____ and _____. Each organ is made of layers called _____. Each tissue is made up of _____. The cells in a tissue are _____ and they do the same _____.

2 Write these in order of size, with the largest first.

human   skin cell   skin   outer layer of skin

For your notes:

- Animals and plants are made up of **organs**. Each organ has a job to do.

- Organs are made up of **tissues**.

- Tissues are made up of **cells**. The cells in a tissue are alike.

A2 Building blocks

More about cells

In 1665 a scientist called Robert Hooke used a microscope to look at plants. He saw many tiny boxes. He was the first to call these boxes 'cells'. The word 'cell' means a small room.

All living things are made up of cells. There are two main types of cell: **animal cells** and **plant cells**.

Animal cells

An animal cell is shown here.

Robert Hooke.

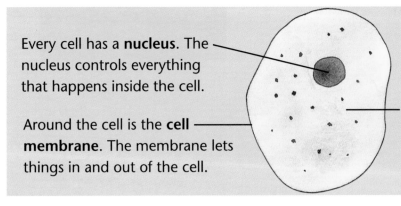

Every cell has a **nucleus**. The nucleus controls everything that happens inside the cell.

Around the cell is the **cell membrane**. The membrane lets things in and out of the cell.

Inside the cell is a liquid called the **cytoplasm**. This is where **chemical changes** happen to keep the cell alive.

a What is the job of: (i) the cell membrane? (ii) the nucleus?

b What is the name of the liquid inside the cell?

Plant cells

A plant cell is shown here. It has a cell membrane, a nucleus and cytoplasm, like an animal cell. Plant cells have some extra parts as well.

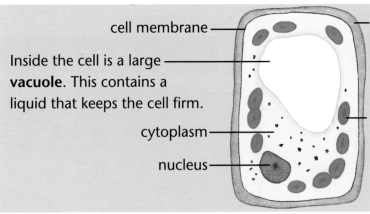

cell membrane

Inside the cell is a large **vacuole**. This contains a liquid that keeps the cell firm.

cytoplasm

nucleus

Every plant cell has a **cell wall**. This supports the cell and makes it strong. It is like a box. The cell wall is made from **cellulose**.

Many plant cells have **chloroplasts**. These contain a green substance called **chlorophyll**. Plants make their food in the chloroplasts. The chlorophyll makes the plants green.

c What is the cell wall made from?

d Why do most plants look green?

A whole cell

When we look at cells through a microscope we see only a slice through a cell. The pictures on this page show slices like this. It is like taking a very thin slice through a celery stick.

Cells are really more like tiny bricks. This picture below shows a whole cell and a slice through one.

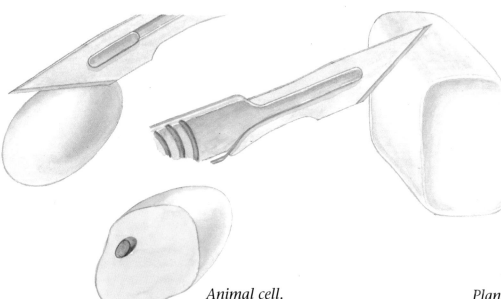

Animal cell.

Plant cell.

Questions

1 Copy and complete these sentences using the words below.

**cell wall   cell membrane   chloroplasts   cytoplasm
nucleus   vacuole**

All cells have a _____ _____, _____ and a _____.
Plant cells also have a _____ _____ and a large _____.
Green plant cells also have _____.

2 Copy this list of parts of a cell. Write 'P' if a plant cell has it. Write 'A' if an animal cell has it.

   a cell membrane **b** cell wall **c** cytoplasm **d** nucleus
   e chloroplasts **f** vacuole.

3 Which part of a cell does each job below?

   a controls the cell

   b lets stuff in and out of the cell

   c place where chemical changes happen.

4 Imagine you are Robert Hooke. Write a letter to a friend describing your microscope and what it does.

For your notes:

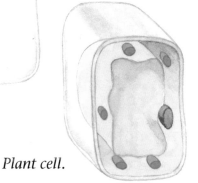

- All living things are made of cells.

- There are two types of cell: **animal cells** and **plant cells**.

- Both types of cell have a **cell membrane, cytoplasm** and a **nucleus**.

- Plant cells also have a **cell wall** and a **vacuole**. Many plant cells have **chloroplasts**.

A3 Cells and growth

How living things grow

Living things start small and get bigger. This is called **growth**. You started off life as a single cell the size of the full stop at the end of this sentence. You have grown a lot since then!

This growth happens by:

● making more cells

● cells getting bigger.

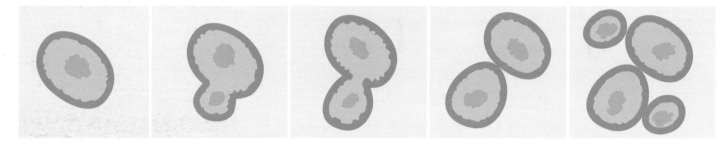

Making more cells

Cells divide. One cell becomes two cells. This is called **cell division**. The pictures below show one yeast cell making two and then four cells.

a Look at the pictures above. How many times has the yeast cell split into two?

b What is the scientific name for this splitting?

Cell division also happens in plants and animals. The photos below show a plant cell dividing.

These photos show an animal cell dividing. The cells have been stained with dyes. This makes the parts of the cell show up better.

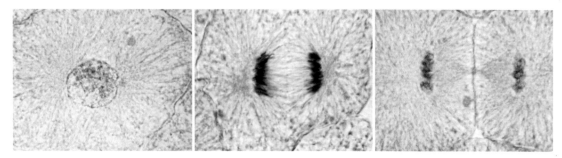

Look back at the photos of a plant and animal cell dividing. When a cell divides, the nucleus divides first. This is because the nucleus contains the information that controls the cell.

Cells getting bigger

When a cell divides, it makes two smaller cells. Look again at the cells in the photos. The new cells are smaller. The new cells will increase in size and then divide again.

You can also see this with the yeast cells. Look at photos of the yeast cells dividing. The new cell starts as a little bud and then increases in size.

c What would happen if a cell did not get bigger between divisions?

Did you know?

Humans start life as a single cell. An adult human contains about 50 000 000 000 000 cells. A single cell would have to divide 46 times to make enough cells for an adult human

Questions

1 Copy and complete these sentences.

All cells are made from …

Cells split to make other cells. This is called …

Growth happens when more cells are made and the cells …

2 **a** How many cells are made when a cell divides?

b Which part of the cell divides first?

c Are new cells bigger or smaller than the cell that divided?

d What happens to new cells before they divide again?

3 Look at the three pictures below. Write the letters in the correct order to show cell division.

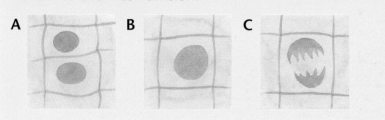

For your notes:

- All cells are made from other cells.

- Cells divide into two to make more cells. This is called **cell division**. The nucleus divides first.

- Cells increase in size after they divide.

- **Growth** happens by cells making more cells and cells getting bigger.

A4 Scaling up and down

Too small to see

A cell is smaller than the end of a pencil, so it is impossible to draw a cell at its real size. We need to draw it much larger than it really is. Scientists call this a **scale diagram**.

Scale diagrams are also useful for showing things that are too big to fit on a page. We draw big things smaller than they really are. Maps are scale diagrams.

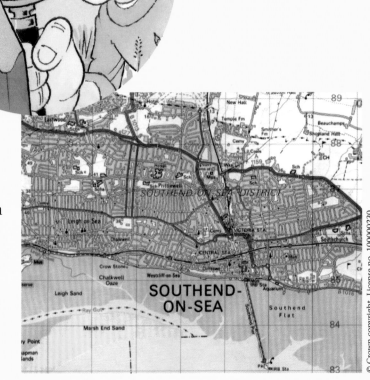

ⓐ Discuss in your group other uses for scale diagrams.

Scales

Scaling up means showing an object bigger than it really is. **Scaling down** means showing an object smaller than it really is.

The picture shows Mr and Mrs Beetroot with their two children Nick and Aileen. They are not really this small. They have been scaled down. To find out how big they really are, we have to scale up again.

In this picture, 1 cm is used to show 40 cm in real life. In the picture Mr Beetroot is 5 cm tall. To find his real height you multiply his height in the picture by 40. This means you scale up by 40.

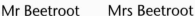

> 5 x 40 = 200 So Mr Beetroot is 200 cm tall.

Instead of using lots of words to describe scaling up, we can say the **scale factor** for the picture is 40. This means you multiply by 40. So 1 cm represents 40 cm.

Mr Beetroot   Mrs Beetroot   Nick   Aileen

ⓑ Copy and complete the table to find the real heights of the rest of the Beetroot family.

Name	Picture height in cm	Scale factor	Picture height x scale factor	Real height in cm
Mr Beetroot	5	40	5 × 40	200
Mrs Beetroot	4	40		
Nick	3	40		
Aileen	2	40		

When we want to scale up, we multiply by a scale factor. Now imagine we want to scale down.

c Discuss in your group how you think we can scale down using the scale factor.

If you want to draw Mr Beetroot scaled down by a scale factor of 20, you divide his real height by 20.

200 ÷ 20 = 10 cm   So you would draw Mr Beetroot 10 cm tall.

d Draw a table with these headings and work out how big you would draw the rest of the Beetroot family.

Name	Real height in cm	Scale factor	Real height ÷ scale factor	Picture height in cm
Mr Beetroot	200	20	200 ÷ 20	10

Questions

1 The following table lists some of the objects in the Beetroots' house.

a Copy and complete this table by scaling up.

Object	Real measurement in cm	Picture measurement in cm	Scale factor
length of car		10	30
length of pencil		2	10

b Copy and complete this table by scaling down.

Object	Real measurement in cm	Picture measurement in cm	Scale factor
height of car	150		30
height of bicycle	80		20

2 Measure your height in centimetres. Draw a line to show your height scaled down by a factor of 10.

A5 Flower cells

Parts of a flower

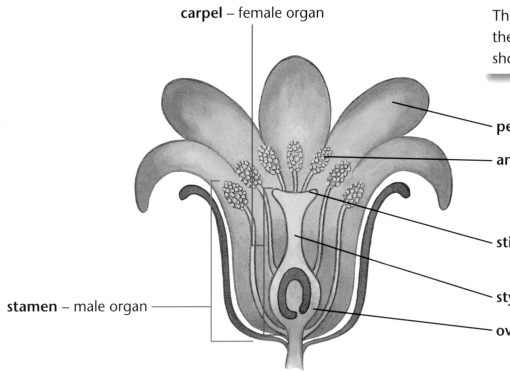

carpel – female organ

petal – attracts bees

anther – part of the stamen that makes pollen grains

stigma – where the pollen grains land

style – holds up the stigma

ovary – makes egg cells

stamen – male organ

Do you remember?

The parts of the flower and the job each part does are shown in this picture.

Pollination

The job of a flower is **reproduction**, making more plants.

The male organs in flowers make **pollen grains**, the male sex cells. You can see pollen grains in this photo.

The female organs in flowers make **egg cells**, the female sex cells.

The pollen grains from an anther on one plant have to be transferred to the stigma on another plant. This is called **pollination**.

a What is the job of:
   (i) the anther?   (ii) the ovary?

b What is pollination?

c Which insect often helps pollinate flowers?

Fertilisation

Pollination puts the pollen grain on the stigma. The pollen grain then grows a **pollen tube**. You can see this in the picture. The nucleus from the pollen grain travels down the pollen tube to the ovary.

Fertilisation happens when the nucleus of the male sex cell joins with the nucleus of the female sex cell. This makes a fertilised egg cell.

The ovule, with the fertilised egg cell inside it, makes a **seed**. The fertilised egg cell will grow into a new plant.

d What grows from the stigma to the ovary?

e Where does fertilisation happen?

f What two things join during fertilisation?

1 A pollen tube forms and grows down the style, carrying the nucleus of the pollen grain.

2 The pollen tube grows into the ovary, then the ovule.

3 The nucleus of the pollen grain joins with the nucleus of the egg cell and fertilises it.

pollen grain
stigma
style
pollen tube
ovule
egg cell
ovary
pollen grain nucleus

Questions

1 Pair up each part of the flower with its job (function).

Parts	Jobs
anther	makes eggs
ovary	attracts bees
petal	holds up stigma
stigma	landing platform for pollen
style	makes pollen grains

2 How do pollen grains get from one flower to another?

3 How does a pollen grain get its nucleus from the stigma to the ovary?

For your notes:

- Flowers contain the sex organs of the plant.

- The male sex cells are the **pollen grains**.

- The female sex cells are the **egg cells**.

- **Pollination** is the transfer of pollen grains from the **anther** to the **stigma**.

- **Fertilisation** happens when the nucleus of a pollen grain joins with the nucleus of an egg cell.

B1 Spot the difference

Why reproduce?

Do you remember?

Wind and insects carry pollen from one flower to another. This way the flower can make seeds, and the seeds grow into new plants.

Just like plants, animals need to reproduce or they will die out. Giant pandas are becoming very rare in the wild and zoos all over the world are trying to breed them so that they will not become extinct.

More of the same

Most animals need a female and a male to reproduce. Humans have children that grow into adults. The bodies of men and women are different so that they can produce children. Men have a male reproductive system and women have a female reproductive system.

Men make sex cells called **sperm**. Women make sex cells called **eggs**. To make a baby, a sperm and an egg must join together.

ⓐ **Why are the bodies of men and women different?**

Male reproductive system

The man's reproductive system makes sperm. It also makes it possible for the sperm to get into the woman's reproductive system.

The picture on the right shows the male reproductive system from the side.

Sperm are made in the **testes**. As they pass along the **sperm tube**, glands add a liquid to the sperm to make **semen**. The semen leaves the man's body through the **penis**.

ⓑ **Where are sperm made?**

ⓒ **List the parts, in order, that the sperm go through on their way out of the man's body.**

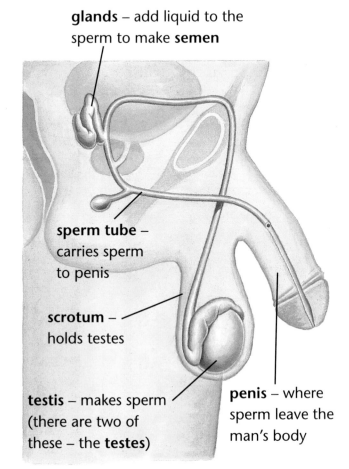

glands – add liquid to the sperm to make **semen**

sperm tube – carries sperm to penis

scrotum – holds testes

testis – makes sperm (there are two of these – the **testes**)

penis – where sperm leave the man's body

Female reproductive system

The woman's reproductive system makes eggs. It has a place where a sperm and an egg can join. It also has a place where the new baby can grow.

The picture shows the female reproductive system from the front.

There are two **ovaries**, one on each side. Once a month an egg leaves one of the ovaries and passes down the **oviduct** (egg tube).

The **uterus** (womb) is where the baby develops.

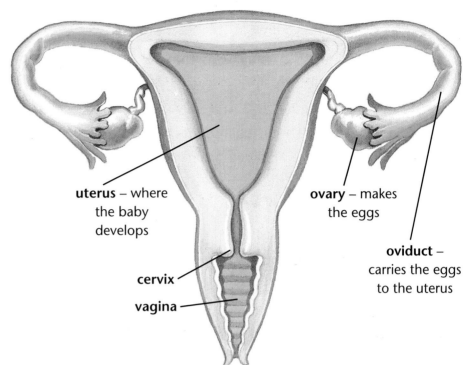

uterus – where the baby develops

ovary – makes the eggs

oviduct – carries the eggs to the uterus

cervix

vagina

The opening of the uterus is called the **cervix**. This is a ring of muscle that opens wide enough to let the baby out at birth.

When it is born the baby passes through the **vagina**.

d Where are the eggs made?

e Where does an egg go after it has left an ovary?

f What happens in the uterus?

Questions

1 Copy and complete these sentences using the words below.

**ovaries   testes   semen   uterus
oviduct   penis   sperm tube**

a Sperm are made in the _____. When the sperm leave the testes, they pass down the _____. Glands add a special liquid to make _____. The sperm then leave through the _____.

b Eggs are made in the _____. Every month an egg leaves one of the ovaries and passes down the _____ to the _____.

2 How many eggs does a woman usually release in one year?

3 Why is the cervix made of muscle?

For your notes:

- Just like plants, animals need to reproduce or they will die out. To do this they have male and female parts.

- Men make **sperm** in the **testes**. The sperm pass along the **sperm tube** and out of the **penis**.

- Women make **eggs** in the **ovaries**. The eggs pass along the **oviduct** to the **uterus**.

Sperm and egg cells

The photos show sperm cells and an egg cell seen under a microscope.

Sperm and egg cells have a nucleus, cell membrane and cytoplasm like any other cell. They have special features to help them do their different jobs better.

Look at the top photo. Sperm cells have long tails to help them swim to the egg.

Look at the bottom photo. Egg cells are round and much bigger than sperm cells. They have a special outside layer which stops more than one sperm getting in.

How sperm and egg meet

To make a baby, the male and female sex cells must meet and join together. When a man and a woman make love, the man's penis enters the woman's vagina. This is called **sexual intercourse**. Sperm leave the penis and swim into the vagina. The sperm then swim through the uterus and into both oviducts.

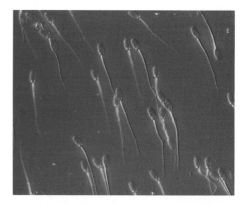

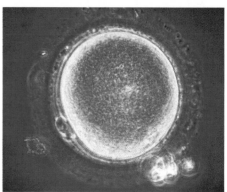

Did you know?

During sexual intercourse, up to 500 million sperm are released into the vagina.

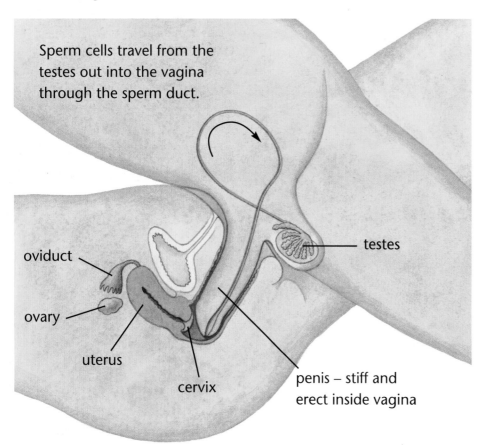

Sperm cells travel from the testes out into the vagina through the sperm duct.

oviduct

ovary

uterus

cervix

testes

penis – stiff and erect inside vagina

(a) What two cells must meet to make a baby?

(b) How do the sperm get into the woman's body?

What happens next?

What happens next depends on whether an egg has been released into the oviduct.

2 The egg passes along the oviduct.

3 A sperm meets the egg in the oviduct and fertilises it.

4 The fertilised egg passes along the oviduct to the uterus.

oviduct

ovary

uterus

vagina

1 An egg is released into the oviduct. Sperm are released into the vagina during sexual intercourse. They swim up through the uterus.

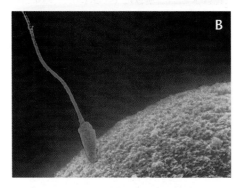

A

B

If there is an egg in the oviduct:

- the sperm will surround it, as shown in photo A.

- one sperm burrows into the egg. Photo B shows this.

- the nucleus of the sperm joins with the nucleus of the egg. This is called **fertilisation**.

- the fertilised egg will become a baby.

If there is no egg in the oviduct all the sperm will soon die. No baby will be produced.

C **What happens when a sperm meets an egg in the oviduct?**

For your notes:

- Sperm and egg cells are adapted to help them do their different jobs better.

- In **sexual intercourse**, millions of sperm are released into the woman's vagina. One might fertilise an egg.

- **Fertilisation** happens when the nucleus of a sperm joins with the nucleus of an egg.

Questions

1 Julie made some cards to show the stages in fertilisation. Write down the statements on the cards in the correct order.

| The nucleus of the sperm joins with the nucleus of the egg. | One sperm burrows into the egg. | Sperm surround the egg. |

2 Why do you think so many sperm are produced?

3 Explain what fertilisation means.

From egg to baby

At fertilisation a new cell is made. It divides into two. These two cells then divide into four, and the four cells divide into eight, and so on. The tiny ball of cells is called the **embryo**.

a **What is an embryo?**

The embryo settles into the lining of the uterus. The woman is now **pregnant**.

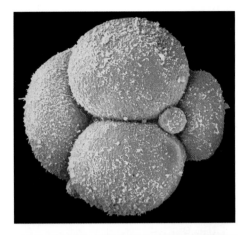

The fertilised egg passes along the oviduct into the uterus.

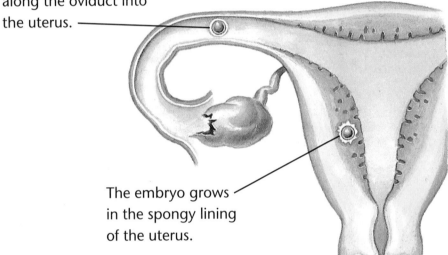

The embryo grows in the spongy lining of the uterus.

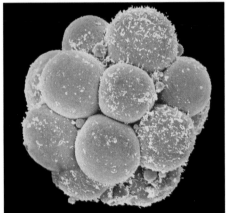

The growing baby

The embryo grows more to become a **fetus**. This is a developing baby. The photos below show the development of the fetus during **pregnancy**.

 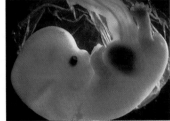  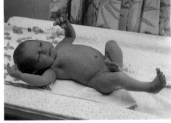

An embryo at 4 weeks.   A fetus at 9 weeks.   A fetus at 22 weeks.   At around 39 weeks the baby is born.

Getting what it needs

The fetus gets the food and oxygen it needs from the mother. It does this through the **placenta** and the **cord**.

The fetus gets rid of carbon dioxide and other waste substances through the placenta.

The blood in the cord carries food and oxygen to the fetus from the placenta. Blood carries the carbon dioxide and waste substances back.

ⓑ **What links the baby to the placenta?**

ⓒ **What is the job of the placenta?**

Birth

Pregnancy lasts for about nine months. Then the baby is ready to be born. It is pushed out of the uterus by **contractions**. These happen when the strong muscles of the uterus wall squeeze.

After it is born, the baby is still attached to the mother by the cord. The cord has to be cut and tied. The placenta is no longer needed and leaves the uterus a few minutes later as **afterbirth**.

The mother produces milk as food for the baby after it is born. Breast milk is very nutritious. It contains substances that protect the baby from catching diseases.

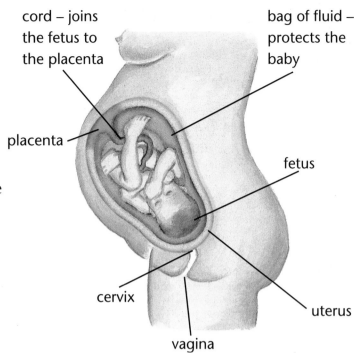

cord – joins the fetus to the placenta

bag of fluid – protects the baby

placenta

fetus

cervix

vagina

uterus

ⓓ **How long is pregnancy?**

ⓔ **What is the afterbirth?**

Questions

1 Write out each part along with its correct job.

Parts	Jobs
cord	supplies the fetus with food and oxygen
placenta	joins the placenta to the fetus
afterbirth	pushed out of the uterus after the baby is born

2 Copy and complete these sentences.

Pregnancy lasts about _____ months. The baby is pushed out by the strong _____ of the uterus. The _____ that joins the baby to the placenta is cut and tied when the baby is born. The _____ also comes out of the uterus a few minutes later. This is known as the _____.

3 Make a leaflet for Year 6 pupils describing how the baby develops inside the mother and is born after nine months. Look at the pictures and text on these pages to help you.

For your notes:

- It takes nine months for a human baby to develop fully inside its mother. This is called **pregnancy**.

- The **fetus** is joined to the mother by the **cord** and the **placenta**.

- At birth, the baby is pushed out of the uterus by strong **contractions**. The placenta then leaves the uterus as **afterbirth**.

B4 The menstrual cycle

What causes periods?

As we grow up physical changes happen in our bodies. The changes make it possible for us to have babies.

At around 10 to 14 years girls start a monthly cycle called the **menstrual cycle**. The cycle lasts about 28 days. It is controlled by substances called **hormones**.

These hormones cause an egg to develop and be released in each cycle. The lining of the uterus builds up.

If the egg is fertilised, it settles into this lining. If the egg is not fertilised, it dies. The lining of dead cells and blood breaks down and leaves the body through the vagina. This is known as a **period**. We say that the period starts on day 1 of the cycle.

The diagram shows the menstrual cycle.

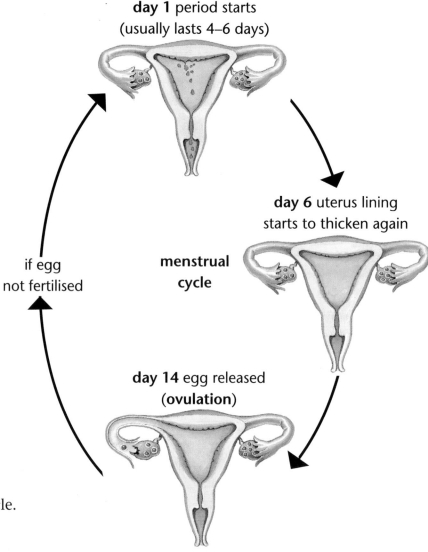

day 1 period starts
(usually lasts 4–6 days)

day 6 uterus lining
starts to thicken again

menstrual cycle

if egg not fertilised

day 14 egg released
(**ovulation**)

ⓐ **What is a period?**

ⓑ **Look at the picture, in a 28-day menstrual cycle, on what day will the egg be released?**

When a girl or a woman's period finishes, a new egg will develop in the ovary and the cycle continues.

Becoming pregnant

If the woman has sexual intercourse between days 13 and 15, and any sperm reach the oviduct, one can fertilise the egg when it is released. If the egg is fertilised, the uterus lining is needed to protect the embryo. The lining does not break down, so periods stop during pregnancy.

ⓒ **What happens to a woman's periods during pregnancy?**

When will it happen?

Kellie started her periods in March during Year 7. She marked the date with a circle in her diary. She marked the first day of her next two periods with a cross.

d How long is Kellie's cycle?

One at a time

Usually, only one egg is released and fertilised at a time. This is because the human reproductive system is designed to make one baby at a time.

Twins

Sometimes a woman gives birth to more than one baby at the same time. Two babies together are called **twins**.

Identical twins, like Hannah and Mary, come from just one egg. The egg splits into two just after fertilisation. Because both Hannah and Mary come from the same egg and sperm, they look exactly the same.

Non-identical twins, like Charlie and Amy, are made if two eggs leave the ovaries at the same time. Each egg is then fertilised by a different sperm. These twins are as different as other brothers and sisters.

e Give one difference in the way identical twins and non-identical twins are made.

For your notes:

- The **menstrual cycle** is a monthly cycle in women, controlled by **hormones**. During the cycle an egg is released, and the woman has a **period**.

- **Identical** twins come from one egg that splits into two after fertilisation.

- **Non-identical** twins are made when each egg is fertilised by a different sperm.

Questions

1 Copy the table. Complete it to show what happens at each stage of a 28-day menstrual cycle. The first one has been done for you.

Day 1	period starts (usually lasts 4 to 6 days)
Day 6	
Day 14	

2 Explain why only one egg is usually released and fertilised at a time.

3 Suggest two reasons why a woman's periods stop when she is pregnant.

B5 Adolescence

Take good care

Human babies need a lot more care from their parents than other newborn mammals. They cannot feed themselves or walk when they are born.

But baby zebras can run around soon after they are born. They need to do this to avoid being eaten by other animals. Humans look after their young until they are adult, so they have a good chance of surviving!

All change!

You are at the **adolescence** stage. Adolescence is a time in everyone's life when lots of physical and emotional changes happen. The changes prepare us to be young adults. The changes happen at different times in different people.

Puberty is the first part of adolescence. Most of the physical changes take place during puberty. These changes make it possible for us to have babies. Puberty usually starts earlier in girls than it does in boys.

In puberty, young people often find that their emotions and behaviour change. They become more attracted to the opposite sex.

Adolescence finishes when people stop growing, at about 18 years.

ⓐ **What happens in adolescence?**

ⓑ **What is the first part of adolescence called?**

During puberty

Many changes happen to boys and girls in puberty. Hormones are substances that make these changes happen.

The changes that happen in puberty are given in the table on the next page.

Some are also shown in the pictures in the table.

Changes in boys	Changes in girls
sudden increase in height (growth spurt)	sudden increase in height (growth spurt)
hair starts to grow on body, including pubic hair	hair starts to grow on body, including pubic hair
voice deepens	breasts grow
testes start to make sperm and hormones	ovaries start to release eggs and make hormones
shoulders broaden	hips widen
sexual organs get bigger	periods start

boy → man

girl → woman

c What is the job of the testes?

d What happens in the ovaries at puberty?

Size matters

When a person grows, their cells split into two. This happens again and again, to make more cells. This is cell division. At first there is no increase in mass because a big cell divides into lots of smaller new cells. Then the cells increase in size, so the mass increases. This is **growth**.

Growth spurts

A growth spurt is a time of rapid growth. It happens when cells divide rapidly and get bigger. The fetus has a growth spurt. Adolescence is another time when we have a growth spurt.

e When do growth spurts happen?

f Why do you think that changes happen to our bodies in puberty?

Questions

1 Write out each word with its correct explanation.

Words	Explanations
growth	a time in everyone's life when physical and emotional changes take place
puberty	the first part of adolescence, when most of these changes take place
adolescence	substances that cause the changes in boys and girls
hormones	cells divide into two and the new cells increase in size

2 **a** Describe three changes that happen to boys during puberty.

b Describe three changes that happen to girls during puberty.

3 If an egg is not fertilised, what happens next in the menstrual cycle?

For your notes:

- Humans care for their babies more than other mammals do.

- **Adolescence** is a time when physical and emotional changes happen.

- **Puberty** is the first part of adolescence, when most of the physical changes happen.

- **Growth** happens when cells divide and increase in size.

Think about:

B6 Pregnant pause

Time to develop

The length of time that an animal is pregnant is called its **gestation period**. It is the time from fertilisation to birth.

A mouse is pregnant for only three weeks. An elephant is pregnant for nearly two years!

The table shows some data about the length of time that different animals are pregnant. The animals are in order of size, with the smallest animal first.

Animal	Gestation period in days	Animal	Gestation period in days
mouse	21	human	280
squirrel	30	camel	355
cat	62	rhino	420
kangaroo	40	elephant	649
ape	200		

Did you know?

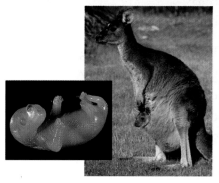

Kangaroos have a very short gestation period. They give birth after about five weeks, but the baby is still much too helpless to survive on its own. It crawls into its mother's pouch and stays in there for another six months. The koala and wallaby also do this.

To help you see patterns in this data more clearly, you can turn it into a bar chart. The first three rows have been done for you below.

ⓐ Copy this bar chart and add bars for all the other animals in the table.

ⓑ Which is the biggest animal?

ⓒ Which is the smallest animal?

ⓓ Which animal has the longest pregnancy?

ⓔ Which animal has the shortest pregnancy?

Look carefully at your chart. Can you see a pattern in your results?

ⓕ Altaf thinks that there is no relationship between the size of an animal and its gestation period. Do you agree with him? Explain your answer.

ⓖ Look at your chart again. Is there any animal that doesn't fit the pattern?

ⓗ What do you know about this animal that might explain this?

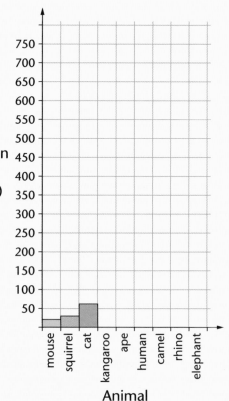

Gestation period (in days)

Predators and prey

Another way of looking at data is to use a line graph. On the right is some data for three more animals. They have all been put in order of their mass. They are all **predators**. This means that they hunt and eat other animals – their **prey**.

Animal	Average adult mass in kg	Gestation period in days
cheetah	95	60
lion	190	108
tiger	210	109

We can draw a line graph of this data. Look at the graph on the right, we put animal mass along the bottom and gestation period up the side. For the cheetah, for example, we put a point where a straight line up from 95 on the bottom meets a straight line coming across from 60 on the side.

The next table shows data for three prey animals.

Animal	Average adult mass in kg	Gestation period in days
antelope	45	180
wildebeest	200	255
zebra	280	360

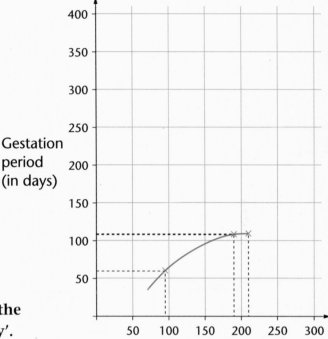

Gestation period (in days)

Average adult mass (in kg)

i Copy the graph and plot this data on it. Join the points with a curved line. Label this line 'prey'.

j An impala has an average adult mass of 55 kg. What do you think its gestation period will be?

k You can predict the gestation period for a different animal if you know its average adult mass. Draw a line up from the bottom until it touches the line graph, then draw a line to the side and read the gestation period.

Which line is higher on your graph, the predators or the prey?

l What does this tell you?

When a lion cub is born, it is blind and helpless for a week or so. It cannot move very far, and its mother has to look after it. A baby zebra is very different. It can walk with the rest of the herd only a few hours after it is born.

The longer a baby animal spends inside its mother, the more developed it will be.

m A zebra spends a lot longer developing inside its mother than a lion does. How does this help the zebra?

23

C1 Environments

The environment

The **environment** is the world around us. Sunlight, water, soil, rock, air and other living things are all part of the environment.

● A desert is an environment that gets very little rainfall. This means it is very dry.

● A rain forest gets a lot of rainfall and is very wet.

● Polar environments (the Arctic and Antarctic) are very cold.

● Ponds, sea, beaches, woodlands and grasslands are other environments.

Do you remember?

An **organism** is a living thing, such as a plant or animal. A **habitat** is where a living thing lives. The '**conditions**' in a habitat mean what it is like, such as hot, wet, dark, light, etc.

A

B

C

(a) Photos A, B and C show three different environments: desert, polar and rain forest. Which photo is which environment?

Living in the desert

Many plants and animals live in desert habitats, like photo C. Look at the photo on the right. It shows a saguaro cactus. This cactus has many features that help it survive in the desert.

These special features are called **adaptations**. We say that the cactus is **adapted** to the desert.

It does not have leaves, so it loses less water by evaporation.

Its roots grow outwards 30 metres in all directions to get as much rain water as possible.

It can store a lot of water in its stem.

Saguaro cactus.

Its roots stay very close to the surface to get water when it rains.

(b) Plants take in water through their roots. They lose a lot of water from their leaves. Explain how a saguaro cactus survives in the desert.

Animals living in the desert also have adaptations. The kangaroo rat is a very successful desert animal. It has adaptations to:

- stop it losing water

- avoid very high temperatures

c Explain how the kangaroo rat is adapted to not lose water.

d How is it adapted to avoid high temperatures?

e How is the Kangaroo rat adapted to move around in the desert?

It never comes out during the heat of the day.

It has strong back legs so that it can jump up to 2.75 metres.

Kangaroo rat.

Its toes are wide-spread to stop it sinking into the sand.

It loses very little water in its sweat and urine.

Its feet are hairy for a good grip on the sand. This helps it change direction.

It puts bits of cactus over the opening of the burrow to keep the moist air in.

It digs a burrow to live in. The sand deeper down is cooler and more moist.

Questions

1 Copy and complete these sentences using the words below.

environment   adapted   desert   polar   rain forest

Our _____ is the world around us. _____ environments are very cold. _____ environments are very dry. _____ environments are very wet.

Living things are _____ to their environments. This helps them survive.

2 Look at photos **D** and **E**. One shows a desert fox. The other shows an arctic fox.

a Which photo shows the fox from the Arctic?

b Why is one fox white and the other fox brown?

c Why does one fox have a much thicker coat?

For your notes:

- A living thing's **environment** is everything that surrounds it, including its **habitat**.

- Living things are **adapted** to the **conditions** in their habitat. This helps them survive.

C2 A day in the life of...

Day and night

Think about a wood. During the day it is brighter and warmer. During the night it is darker and colder. Some woodland animals, like squirrels, are active during the day and sleep at night. Other woodland animals, like owls, are active during the night and sleep during the day.

a Look at the photo of the owl. What feature helps it hunt at night?

Tides

Another daily change is the tide. Every day the tide comes in and goes out twice. Part of the beach is uncovered at low tide.

The living things that live on the beach have to be adapted to three different environments.

When the tide is in ...	When the tide is changing ...	When the tide is out ...
living things are underwater.	they are pushed and pulled by strong currents.	they are exposed to air and may dry out.

Martin is interested by the seaweed. It is gooey and slimy. It is stuck to the rocks. It is covered in blobs that pop when you squeeze them, like bubble wrap. Martin's dad tells him that the seaweed is called bladderwrack. The blobs are bladders. They are filled with air.

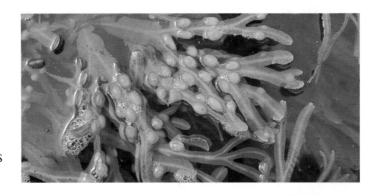

The bladderwrack is adapted to the daily changes that happen on the beach.

When the tide is in …	When the tide is changing …	When the tide is out …
the bladders float so that part of the seaweed stays close to the sunlight. The seaweed traps light to make its food.	the seaweed is stuck to the rock by a holdfast. It cannot be carried away by the current.	the seaweed is covered with a gooey slime. This stops it drying out.

b How is the bladderwrack adapted to cope with the strong currents?

c Why doesn't the bladderwrack dry out at low tide?

d How does the bladderwrack feed?

Barnacles live on the rocks on the beach. Look at the photo on the right. Inside each bony cone there is a little animal.

When the tide is in...	When the tide is changing...	When the tide is out...
the barnacle waves its feathery feet, collecting food.	the barnacle is stuck to the rock by its back.	it closes its bony plates to trap water inside.

e Why doesn't the barnacle dry out at low tide?

f How does the barnacle feed?

g Why isn't the barnacle carried away by the currents?

Questions

1 Copy and complete these sentences. Choose from the words below.

    drier darker brighter warmer wetter colder

During the day, the wood is _____ and _____ than at night.

During low tide the beach is _____ than at high tide.

For your notes:

- Conditions in a habitat change during a day.
- Living things are adapted to the daily changes in their habitat.

C3 Changing seasons

Learn about:
● Adaptation to seasonal change

Winter and summer

Sunlight is less bright and less hot in winter. The days are shorter. This means that it is colder in winter than in summer. Plants and animals need to be adapted to the difficult winter conditions.

Off season for plants

Plants can make very little food in the winter.

ⓐ Why do plants make less food in winter?

Many plants adapt by 'shutting down' in the winter. They become inactive or **dormant**.

Some plants lose their leaves and shut down. Look at the photos below of the oak wood. In the winter, the trees have lost their leaves.

Some plants survive the winter underground. Daffodils survive underground as bulbs.

Other plants survive the winter as seeds. Look at the photos below. Poppy flowers make seeds that survive the winter.

ⓑ Describe three ways plants shut down to survive the winter.

> **Do you remember?**
>
> Green plants need sunlight and water to make their food.

Toughing it out

Winter is a difficult time for animals too. It is cold and food can be hard to find.

Many animals get fatter by eating lots before winter. They also grow thicker fur.

ⓒ How does being fatter help an animal survive winter?

ⓓ How does having more fur in winter help an animal survive?

Some animals, like squirrels, make a store of food for the winter.

Some animals hibernate. **Hibernation** is a deep sleep. The dormouse in this photo on the right hibernates in the winter so it does not have to find food.

Taking a winter break

Other animals leave winter behind. They move to another habitat with better conditions. We call this **migration**. Many birds migrate, including swallows.

Winter white-out

Many animals are adapted to blend into their surroundings. We say they are **camouflaged**.

The natural landscape in Britain is usually a mixture of browns and green. Animals like rabbits and mice are brown because that blends in with their surroundings.

e **Why is being camouflaged good for a rabbit?**

Snow turns the brown landscape white. The camouflage doesn't work any more.

Some animals in snowy habitats can change their coats for winter. This is so they match the snow. Look at the photos. They show an arctic hare in summer and in winter. The white coat helps the hare stay active in winter.

Questions

1 Copy and complete these sentences using the words below.

camouflaged   hibernate   dormant   migrate

Many plants shut down for winter. We say they are _____.

Some animals move to a warmer habitat in winter.
We say they _____.

Some animals go to sleep for the whole winter.
We say they _____.

Some animals change their coats in winter.
This is so they stay _____.

2 Why does an arctic hare have a brown coat in summer and a white coat in winter? (*Hint:* look at the photos on this page to help you.)

3 How do these living things survive the winter?

a oak tree   **b** daffodil   **c** poppy   **d** dormouse   **e** swallow.

For your notes:

● Conditions in a habitat change with the seasons.

● Some living things live through winter by **hibernating, migrating** or becoming **dormant**.

● Other living things have adapations that allow them to stay active throughout winter.

C4 Adapted to feed

Who eats what?

Living things are either **producers** or **consumers**. Producers make their own food using sunlight. Plants are producers. Consumers eat other living things.

Do you remember?

A **predator** is an animal that hunts and feeds on other animals. A **prey** animal is hunted by a predator for food.

Herbivore adaptations

Animals that eat only plants are called **herbivores**. Cattle are herbivores. The cow in the photo has many adaptations for eating leaves and stems.

Other herbivores, like sheep and horses, have similar adaptations.

ⓐ How do cows' teeth help them feed?

ⓑ How is a cow's gut adapted to eating grass?

Carnivore adaptations

Animals that only eat other animals are called **carnivores**. Carnivores like the cheetah have many adaptations that help them catch and kill prey.

ⓒ Why do cheetahs have spots?

ⓓ Why do cheetahs run so fast?

ⓔ How are a cheetah's jaws and teeth adapted to hunting?

Sharp front teeth help the cow to chop grass.

Flat back teeth with ridges help to grind up the grass.

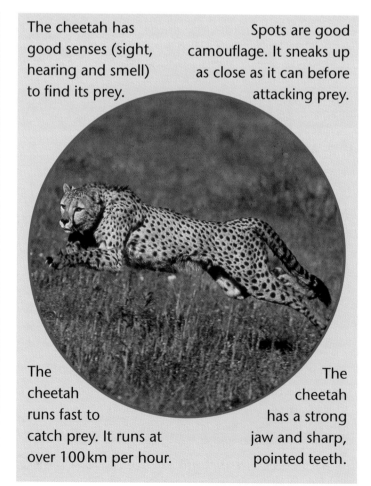

Special microbes in the cow's gut helps it digest leaves and stems.

A cow eats for many hours every day because leaves and stems are low in energy.

The cheetah has good senses (sight, hearing and smell) to find its prey.

Spots are good camouflage. It sneaks up as close as it can before attacking prey.

The cheetah runs fast to catch prey. It runs at over 100 km per hour.

The cheetah has a strong jaw and sharp, pointed teeth.

Prey adaptations

Prey animals, like rabbits, have adaptations to help them escape predators.

The rabbit has strong back legs so it can run fast and dodge.

The rabbit has eyes at the side of the head to help it see predators coming.

The rabbits digs burrows to hide in. It digs with its strong back legs.

Rabbits live and feed in groups. The flash of a white tail is the signal for all the rabbits to run.

Very good hearing helps the rabbit hear predators coming. It has big ears that can turn forwards and backwards, so it can listen in all directions.

Rabbits are camouflaged, so they don't stand out against the brown and green background.

g *Omnivores* are animals that eat both producers and comsumers. Can you think of one omnivore?

Questions

1 Copy and complete these sentences using the words below.

carnivore   herbivore   plants   animals

Animals that eat plants only are called _____. Omnivores are animals that eat _____ and _____. Animals that eat only other animals are called _____.

2 Look again at the photo of the rabbit. How do the following features help the rabbit escape predators?

   a strong back legs

   b all-round eyesight and hearing

   c a brown coat

   d a white tail.

3 Giraffes are herbivores. Why do giraffes have long necks?

For your notes:

- **Producers** make their own food. **Consumers** eat other living things.

- Consumers can be **herbivores, carnivores** or **omnivores**.

- Living things have adaptations so that they are good at getting food.

- **Predators** have adaptations that help them hunt their **prey**.

- Prey animals have adaptations that help them avoid being eaten by predators.

31

Food chains

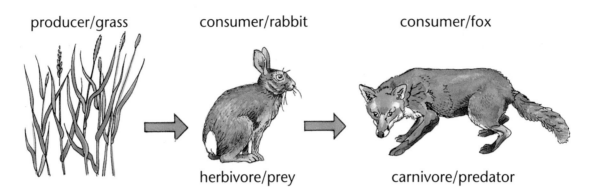

producer/grass   consumer/rabbit   consumer/fox

herbivore/prey   carnivore/predator

As you know, a **food chain** shows you who eats what. In this food chain the grass is the prducer. The rabbit eats the grass and the fox eats the rabbit.

The arrows in a food chain show you how the energy is transferred from producer to consumer.

The energy in a food chain comes from the Sun. The producer (the grass) traps the energy in the sunlight and uses it to make its food. When the rabbit eats the grass, the energy in the plant is transferred to the rabbit.

a **Which of these living things can trap the energy in sunlight?**
● fox   ● grass   ● rabbit

b **Why do food chains always start with a producer?**

c **What do the arrows in a food chain show?**

Interdependence

This food chain tells us that rabbits need grass. If the grass died, the rabbits would die and then the foxes would die.

d **Why would the foxes die?**

Imagine there were no foxes. Fewer rabbits would be eaten. They would breed more and so there would be more rabbits. They would eat all the grass.

e **What would happen when all the grass was gone?**

If the rabbits had no predators they would eat all the grass and starve. Foxes need rabbits but rabbits also need foxes. This is an example of **interdependence**.

> **Did you know?**
>
> Mosses, ferns and seaweeds are producers. Food chains in the sea often start with tiny, one-celled producers. All producers contain the green substance called chlorophyll.

Food webs

There are many food chains in any habitat. These food chains have some of the same things in them. We show this using **food webs**.

The diagram shows a food web for a woodland habitat.

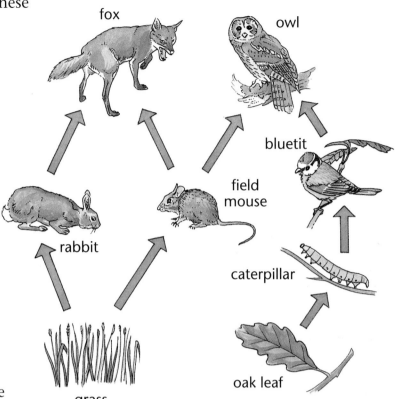

f Find another food chain that starts with grass and ends with fox.

g How many food chains can you find in this food web? (*Hint:* each food chain must start with a producer.)

h Imagine all the fieldmice died. What would happen to the number of:
  (i) foxes?   (iv) grass plants?
  (ii) owls?   (v) bluetits?
  (iii) caterpillars?

Food webs show interdependence much more clearly than separate food chains.

Questions

1 Copy and complete these sentences using the words below.

interdependent chains webs energy producer

Food _____ and food _____ show feeding relationships in a habitat.

Food chains always start with a _____.

The arrows show how _____ moves through the food chain.

The living things in a food web are all _____.

2 a Arrange these three living things into a food chain:

lion grass antelope

b Cheetahs eat antelope. Add a cheetah to your food chain to make a food web.

c Elephants eat grass. Add an elephant to your food web.

d Elephants eat acacia trees. Add an acacia tree to your food web.

e Giraffes eat acacia trees. Add a giraffe to your food web.

f Lions eat young giraffes but not elephants. Show this on your food web with an arrow.

For your notes:

- **Food chains** show feeding relationships. They also show how energy is transferred from producer to consumers, and from consumer to consumer.

- The energy in a food chain comes from the Sun. The producer uses sunlight to make its food.

- **Food webs** show all the food chains in the same habitat.

- All living things in a habitat are **interdependent**.

C6 Grouping living things

Grouping

Scientists put things into groups to make it easier to talk and think about them. Sabrina and Nel are grouping the living things on the opposite page. They decide to group the living things by colour.

Green	Not-green
gut weed	dahlia anemone
oak	coral weed
bush cricket	bladderwrack
cushion star	elephant

Read the information on the opposite page. Then answer these questions about Nel and Sabrina's groups.

ⓐ Are all the living things in the 'green' column producers?

ⓑ Have they put all the producers in the photos in the 'green' column?

ⓒ Look at the 'green' group. Are these living things similar to each other? Give reasons for your answer.

ⓓ Look at the 'not-green' group. Are these living things similar to each other? Give reasons for your answer.

Feeding relationships

Look again at the information on the opposite page.

ⓔ Divide the living things into 'producers' and 'not-producers'.

ⓕ Divide the 'not-producers' into herbivores, omnivores and carnivores.

All producers are green. Let's divide them into 'green' and 'not-green' groups.

I think we should be using the information about how the living things feed.

Dahlia anemone.
About 5 cm high. Lives stuck to a rock. 80 short tentacles. Hunts other animals for food.

Coral weed.
Flat, thin, long (6–7 cm), red fronds. Lives stuck to a rock. Traps sunlight to make food.

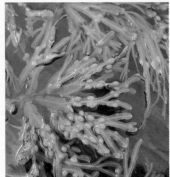

Bladderwrack.
Long, flat, wide (2–5 cm) brown fronds with bladders. Lives stuck to a rock. Traps sunlight to make food.

Great green bush cricket.
4–5 cm long. Hops and flies. Feeds on plants and other insects.

Gut weed.
Thin (about 5 mm wide), green tubes. Lives stuck to a rock. Traps sunlight to make food.

Cushion star.
Up to 7 cm across. Moves across the rocks in the sea. Feeds on other animals.

Oak.
Up to 45 m tall. Green leaves in spring and summer. Lives stuck in the ground. Traps sunlight to make food.

African elephant.
6–7 m long (including trunk), 3–4 m tall (to shoulder). Moves about. Feeds on grass and other plants.

Questions

1 a Group the living things into 'moving' and 'staying in one place'. Make a table showing your two groups.

 b Does this help separate producers from consumers? Which living things are in the 'wrong' place?

 c Nel finds out that sea anemones move about when they are young.

 (i) Which group will you put the dahlia anemone in: 'moving' or 'staying in one place'?

 (ii) Explain the reasons for your decision.

2 Do you agree or disagree with these statements? Give your reasons.

 a 'Green' means producer.

 b 'Green' and 'not moving' means producer.

 c All producers are green and do not move.

Variation and classification

Species

There are millions of living things in the world. To make sense of them all we put them into groups.

Some living things are very similar in the special parts they have or the things they do. These are called their **features**. When living things have a lot of features the same they are in a group called a **species**.

Members of the same species can mate and **reproduce**. Their offspring will also be able to reproduce. Look at the pictures. Wolves and reindeer are both mammals, but different species. They cannot mate to produce a baby 'wolfdeer' or 'reinolf'!

Wolf.

Reindeer.

How do we describe a species?

People describe the features of a species in different ways.

A novel might describe an animal like this:

> Reynard was handsome, proud and independent. … In the autumn, every evening as night was drawing in, he ventured out for his stroll. His elegant russet coat and bold brush blended with the bracken fern and had a particular way of catching the last of the fading light.
>
> That night things went a little differently from normal. He pricked his pointed ears but the sound was not the welcome thud of rabbits' feet …

A field study guide for a biologist might describe the same animal in a much more factual way like this:

- red-brown coat
- bushy tail
- pointed ears
- eats rabbits
- nocturnal
- lives in underground dens

The scientist describes the coat as 'red-brown'. The novel describes the coat as 'russet'. Another writer might choose the word 'rusty'.

Scientists need to describe species and their features very clearly so that other scientists know exactly which species they are talking about. This is why all scientists use the same words, for example 'nocturnal' (awake at night) and 'tail'.

ⓐ Which animal do you think the novel and study guide are describing?

ⓑ How are the two descriptions different?

ⓒ Why is the scientific description more useful for people trying to find the same animal or plant?

The same but different

The humans in this crowd all belong to the same species, because they have many of the same features. But there are differences between them.

We have different colour eyes and hair. We have different weights and heights. Some of us are better at maths than others. Some of us are better at sport.

Differences like these are all examples of **variation** within our species.

ⓓ In what ways are the people in this crowd the same?

No two people are exactly alike – not even identical twins!

ⓔ In what ways are the people in this crowd different?

Questions

1 Copy and complete these sentences using the words below.

   similarities   differences   species   variation

   There are _____ between living things. These differences are called _____. A group of living things with lots of _____ is called a _____.

2 Why do you think that you cannot mate a polar bear with a seal, but you can mate a polar bear with a brown bear?

For your notes:

- If there are enough similar **features** between living things, they are the same **species**.

- Members of the same species can **reproduce** and the species will continue.

- We call the differences between living things **variation**.

Keep it in the family

The members of the Jones family are very alike. So are the members of the Smith family. Each person looks like their parents. This is because some features, like natural eye colour, are passed on or **inherited** from our parents.

You inherited instructions for your eye colour, skin colour and whether you are male or female. They were passed to you in the **sperm** cell from your father and the **egg** cell from your mother.

Amy Smith

Billy Jones

ⓐ What similarities can you see between members of the Jones family?

We are all different

But Jones family members are not all identical. Neither are the Smiths. We are all different because different features are passed on to us from our parents. This is called **inherited variation**.

There are variations between family members because each child in a family inherits a different mixture of features from both parents.

ⓑ What differences can you see between the members of the Smith family?

We are also different because of our surroundings or **environment**. Some of our features are down to choices that depend on our surroundings. Dyeing your hair is an example. This is called **environmental variation**. The pictures show some more examples.

c Look at the pictures above, give two ways we can change our features.

d Give another way in which our surroundings or how we are brought up can affect the way we look.

Questions

1 Copy and complete these sentences by using the words below.

> environment   different   duplicate   identical
> mother   parents   weather

Each individual within a species looks like their _____. But they are not _____ to them. Each sperm and egg has a _____ set of instructions. Differences in the _____ can also cause variation.

2 Donna has these features: a tattoo on her arm, blue eyes, pierced ears, naturally red hair.

Copy and complete the table to classify Donna's features.

Inherited	Environmental

For your notes:

- Some variations between the members of a species are **inherited** from their parents

- Some are caused by their **environment**.

Sorting living things

Living things

We call living things **organisms**. The smallest living things are called **microorganisms** and you need a microscope to see them.

Classifying organisms

There are lots of different species of organism so we put them in groups to make them easier to study. We put species that have similar features into the same group. This grouping is called **classification**. The groups and their descriptions can help us to name any organisms that we find.

The table shows how we start to classify living things. This classification system is used all over the world.

Animals	Plants	Microorganisms	Fungi
● human, horse, spider	● lime tree, primrose	● virus, bacterium	● toadstool, mould
● feed on other animals or plants	● make their own food	● can only be seen with a microscope	● feed on rotting material
● must move around	● green		

ⓐ How do you start to classify living things?

ⓑ Look at the table. How are plants different from animals?

Wherever you look, you will find examples of all these groups. You will find **animals**, **plants**, microorganisms and **fungi** in soil or in a pond. There are many different living things even in very cold places like the Arctic.

Animal X-rays

All animals can be put into two smaller groups, those with a backbone and those without a backbone.

Here are some X-rays of animals from the Arctic.

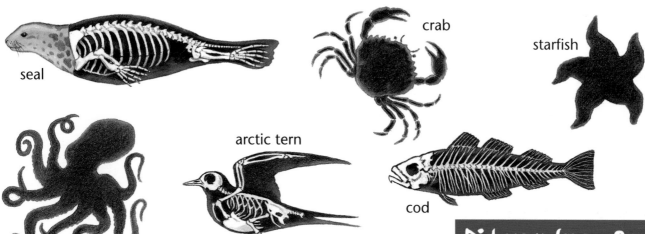

seal

crab

starfish

arctic tern

cod

octopus

C Look at the X-rays. Which of these animals have a backbone?

We call animals with backbones **vertebrates**.

We call animals without backbones **invertebrates**.

Questions

1 Copy and complete these sentences by choosing from the words below.

> **big   bigger   different   groups**
> **organisms   similar   small   smaller**

We can sort species into _____. All of the _____ in a group have _____ features. Each _____ group can be sorted into _____ groups.

2 Copy the table below. Write the following animals in the correct column. You are classifying them.

human   horse   polar bear   octopus   spider   starfish   cod

Vertebrates	Invertebrates

3 Design a key to classify vertebrates and invertebrates.

Did you know?

The first scientist to classify animals was a Frenchman called Georges Cuvier who lived in the eighteenth century.

For your notes:

- We can sort **organisms** into groups with similar features. This is called **classification**.

- **Vertebrates** are animals with a backbone.

- **Invertebrates** are animals without a backbone.

41

D4 More animal groups

Five groups of vertebrates

The vertebrates are divided into five smaller groups:

| ● mammals | ● birds | ● reptiles | ● amphibians | ● fish |

Lion   *Eagle*   *Crocodile*   *Frog*   *Shark*

ⓐ Where do you think humans fit in?

Sort yourself out

Humans have a backbone, so we are vertebrates. We are classified as **mammals**. Lions, apes, dogs, cats and many other furry animals are also mammals.

Mammals

These are the features of mammals:

● Their babies develop inside the mother's body.

● The mother feeds the young on her milk, which she makes in her **mammary glands**.

● They have hairy skin.

The rest of the vertebrates apart from mammals are classified as:

● birds ● reptiles ● amphibians ● fish.

Birds

ⓑ Which features of birds are used for movement?

● **Birds** lay eggs with hard shells.

● They look after their young after they have hatched.

● Birds have feathers and wings.

● Most birds can fly.

Both mammals and birds are warm-blooded. They have ways of keeping their temperature the same.

Did you know?

Mammals and birds are the only groups of vertebrates that look after their young. Reptiles, amphibians and fish usually leave their young to fend for themselves.

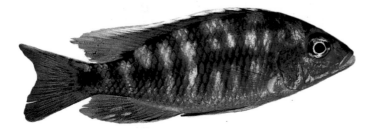

Fish

c How do you think a fish breathes?

- **Fish** can only live in water. They lay eggs in water.

- They breathe through gills.

- They have scales and fins.

Amphibians

- **Amphibians** lay eggs in water. The eggs are like jelly.

- They breathe air and live partly on land, but have to lay their eggs in water.

- They have a smooth, moist skin.

d Why do you think salamanders go back to water in the spring?

Reptiles

e Look at the photo. Describe a crocodile's skin.

- **Reptiles** lay eggs on land. Their eggs have leathery shells.

- They breathe air and live mainly on land.

- They have a scaly, dry skin.

Reptiles, fish and amphibians don't create their own body heat. They are the same temperature as their surroundings.

Questions

1 Copy and complete the table.

Vertebrate group	Features
mammals	have mammary glands, babies develop inside mother's body, have hairy skin
?	lay eggs with hard shells, have feathers and wings
reptiles	?
amphibians	?
?	lay eggs, live in water, breathe through gills, have scales and fins

2 Where do reptiles live and reproduce, on land or in water?

3 Which group of vertebrates feeds its young on milk?

Did you know?

Mammals, birds, reptiles and amphibians breathe air using lungs.

For your notes:

- Vertebrates are classified into five groups.

- The groups are **mammals, birds, reptiles, amphibians** and **fish**.

- Each group has different features.

D5 No bones about it

Invertebrates

The invertebrate animals have no backbones.

We start to classify them by their legs. They have no legs or lots of legs.

No legs

We can start to sort the invertebrates with no legs into groups by the kind of body they have – hard or soft. There are six groups:

● **Jellyfish** have a soft jelly-like body.

● **Flatworms** have a soft flat leaf-shaped body.

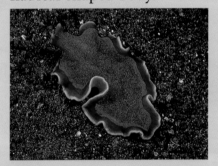

● **Segmented worms** have a soft ringed body.

● **Molluscs** have a soft muscular body with one foot. Most have a hard shell.

● **Starfish** have a hard star-shaped body.

● **Roundworms** have a soft thin round body.

a What feature do we use to classify invertebrates with no legs?

b In which group is an earthworm? It has no legs and a soft body with rings.

Jointed legs

We call the invertebrates with lots of jointed legs **arthropods**. The Latin word for foot is pod. Arthropods have bodies made of sections called **segments**.

We divide the arthropods into four smaller groups:

- **Crustaceans** have a soft body, usually with a hard shell. They have lots of legs.

- **Centipedes** and **millipedes** have a long thin body. They have lots of legs.

- **Spiders** have a 2-part body and 8 legs.

- **Insects** have a 3-part body and 6 legs.

Did you know?

There are three times as many kinds of insect on the Earth as all the other animals put together.

c What is an arthropod?

d What is the difference between a centipede and a roundworm?

Where does it fit?

Trilobites are animals that lived before the dinosaurs. They are arthropods with hard shells. Their bodies have been found preserved in rocks, but many of these are only empty shells.

e Imagine you are the first to find the remains of a trilobite. It has only a hard shell and no body or legs. Look back at the six invertebrate groups with no legs. Which group might you put it in by mistake?

For your notes:

- **Invertebrates** are classified into seven groups.

- The groups are **jellyfish, starfish, flatworms, roundworms, segmented worms, molluscs** and **arthropods**.

- The arthropod group is divided into **crustaceans, centipedes** and **millipedes, spiders** and **insects**.

Questions

1 Copy and complete these sentences by choosing from the words below.

> jellyfish   starfish   flatworm   roundworm
> segmented worm   mollusc   arthropod
> crustacean   spider   insect   millipede   centipede

I have a flat leaf-shaped body and no legs. I am a _____.

I have a star-shaped body. I am a _____.

I have lots of legs and a shell. I am a _____.

2 How can you tell the difference between an insect and a spider?

D6 The right size

Inuit pen pal

Biork lives in Alaska, close to the Arctic Circle.
Biork's ancestors are called Eskimos or Inuit people.

Biork's people have survived the cold Arctic conditions for thousands of years. Biork's grandfather used to go out hunting for seals. He wore clothes made of animal skins and built overnight shelters out of ice to keep warm.

Inuit people are born with short compact bodies. It's a feature that has been passed on through our families.

The Inuit people have short, heavy, compact bodies. Biork has often wondered why she is small. She would like to be tall and thin like her American pen friend in the USA. Biork's grandfather says it's a feature that helps them keep warm.

a　**Why do you think that Inuit are people small?**

Biork's body shape is inherited from her parents, but it also depends on her surroundings, lifestyle and upbringing.

Lots of variables can affect our height and weight.
Some of these are:

● food　　　● exercise

● seasons　● illness

● stress.

Research

A group of Canadian scientists studied the heights of more than 150 Inuit children. They compared them with a **sample** of 150 children in the USA.

The diagram shows what might happen if the sample size is too small.

ⓑ **Why do you think it is better to study a sample of 150 children from each place rather than only 10?**

ⓒ **How would you choose the children that you were going to study?**

Making comparisons

When they had collected their figures, the scientists compared the tallest in each sample and the smallest in each sample. They did this to see the differences.

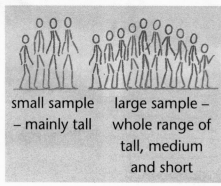

small sample – mainly tall   large sample – whole range of tall, medium and short

Analysing the results

The graph below shows the data for the Inuit girls compared with girls in the USA.

At 12 months old, the lines on the graph for the Inuit children and the American children are close together. This means that their heights are similar. After this, the lines are further apart, showing that the American children are taller.

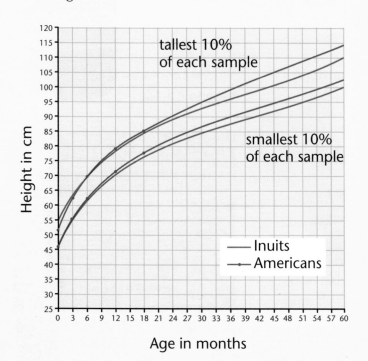

tallest 10% of each sample

smallest 10% of each sample

—— Inuits
—•— Americans

Height in cm

Age in months

Questions

1 Discuss the graph with your partner. Copy and complete these sentences together by choosing from the words below.

   faster   large   slower   small

   There is a _____ difference in the heights of the Inuits and American girls when they are 12 months old. As they get older, the gap widens. The Inuit girls grew _____ than the girls in the USA.

2 What other data should the scientists collect and analyse to check whether the Inuits' small size is inherited? Explain why you made your suggestions.

E1 Acids, bases, alkalis

Acids

An **acid** is a solution of a particular kind of solid or gas in water. An acid that does not contain much water is a **concentrated** acid. An acid that contains quite a lot of water is called a **dilute** acid.

Some acids may be **corrosive**, **toxic**, **harmful** or **irritant**. They have a hazard warning label.

CORROSIVE
may destroy living tissues on contact

TOXIC
poisonous

HARMFUL
may have health risk if breathed in, taken internally or absorbed through skin

IRRITANT
non-corrosive substance which can cause red or blistered skin

(a) **Look at the hazard warning labels. Which hazard do you think is the least dangerous?**

Useful acids

Sulphuric acid is a useful acid. For example, it is used in car batteries. It is corrosive. You need to handle it carefully to avoid burning holes in your clothes or your skin.

Many acids are not corrosive – they do not burn or eat things away. Acids are less corrosive when you dilute them with lots of water.

We use products that contain acids every day. Shampoo, bubble bath and medicines for colds all contain acids.

Vinegar contains ethanoic acid.

Acids in foods

Many of the foods we eat contain acids. They can give food a sour taste.

tannic acid   citric acid   lactic acid   ascorbic acid, which is vitamin C

(b) **Look at these pictures of foods. Make a list of all the acids you can see that are found in foods or drinks.**

ENJOY-ICE COLD
LOW CALORIE SOFT DRINK WITH VEGETABLE EXTRACTS WITH SWEETENER.
INGREDIENTS:
CARBONATED WATER, COLOUR (CARAMEL E 150d), SWEETENER (ASPARTAME), PHOSPHORIC ACID, FLAVOURINGS, CITRIC ACID, PRESERVATIVE (E 211). CONTAINS A SOURCE OF PHENYLALANINE.
Nutrition Information per 100ml
Energy: 1.9kJ, 0.4 kcal
Protein: 0 g
Carbohydrate: 0 g
Fat: 0 g
BEST BEFORE END - see base of can for date
CANNED UNDER AUTHORITY OF THE COCA-COLA COMPANY BY COCA-COLA & SCHWEPPES BEVERAGES LTD., UXBRIDGE, UB8 1EZ

Bases and alkalis

Look at the photo showing substances found in the kitchen or bathroom. They all contain **bases**. A base is the opposite of an acid – it cancels out acidity.

Some bases dissolve in water. We call these **alkalis**. Like acids, many alkalis are corrosive, toxic, harmful or irritant.

c **What is the difference between an alkali and a base?**

A **neutral** solution is neither acidic nor alkaline. Pure water is neutral.

Finding out which is which

Using your senses, such as taste and smell, is not a safe way of finding out whether a solution is acidic or alkaline. The best way is to use an **indicator**. An indicator is a coloured substance that shows whether a solution is an acid or an alkali.

You can make an indicator using coloured dyes from plants. When an indicator is mixed with an acid or alkali, it changes colour. For example, beetroot juice turns pink in acidic solutions and yellow in alkaline solutions.

d **What is the word for a substance that changes colour when it is mixed with an acid or alkali?**

Litmus is an indicator. Litmus paper comes in red or blue paper strips. Acids turn blue litmus paper red. Alkalis turn red litmus paper blue.

Did you know?

Some people use hair relaxers to straighten curly hair. These contain an alkali called sodium hydroxide.

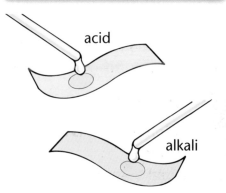

acid

alkali

Questions

1 Copy and complete these sentences by choosing from the words below.

acids   alkalis   corrosive   plants   substances   sour   water

a Acid in food can give it a _____ taste.

b Some acids are dangerous because they are _____.

c Bases are the opposites of _____.

d Indicators are coloured _____.

e Many of them are made from _____.

2 a What colour does blue litmus turn in acid?

b What colour does red litmus turn in alkali?

3 Draw hazard labels for:

a corrosive   b harmful   c irritant.

For your notes:

● **Acids** may be **corrosive, toxic, harmful** or **irritant**.

● **Bases** are the opposite of acids. They cancel out acidity.

● An **alkali** is a soluble base. Alkalis may be corrosive, toxic, harmful or irritant.

● A **neutral** solution is not acidic or alkaline.

● **Indicators** turn different colours with acidic, alkaline and neutral solutions.

Universal indicator

Litmus is very useful for telling whether a substance is acidic or alkaline, but it does not tell you how strong or weak it is or how dangerous it is.

Universal indicator is another indicator made from plants. Because it is a mixture of indicators it gives a range of colours. It will show a different colour for different strengths of acids or alkalis.

● Universal indicator comes in liquid or paper form.

● To use universal indicator liquid, add two drops to the solution you want to test.

● To use universal indicator paper, dip a glass rod into the solution. Touch the indicator paper with the glass rod.

● Compare the colour with the chart.

ⓐ **When might universal indicator be more useful than litmus?**

ⓑ **Why does universal indicator give a range of colours?**

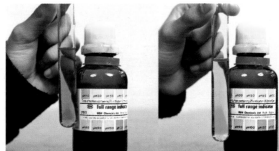

Strongly acidic.

Less strongly acidic.

Weakly acidic.

Neutral.

Weakly alkaline.

Strongly alkaline.

The pH scale

Some acidic solutions are strongly acidic, and others are weakly acidic. The same is true for alkaline solutions. We use pH numbers to measure the strength of the acidity or alkalinity. (*Remember:* write it with a small 'p' and capital 'H'.)

A neutral solution has pH 7. Acidic solutions have a pH less than 7. The lower the pH, the stronger the acidity.

Look at the **pH scale**. Hydrochloric acid is strongly acidic, with pH 1. Vinegar is more weakly acidic, with pH 3–4.

Alkaline solutions have a pH greater than 7. Sodium hydroxide solution is strongly alkaline, with pH 14. Ammonia is more weakly alkaline, with pH 11.

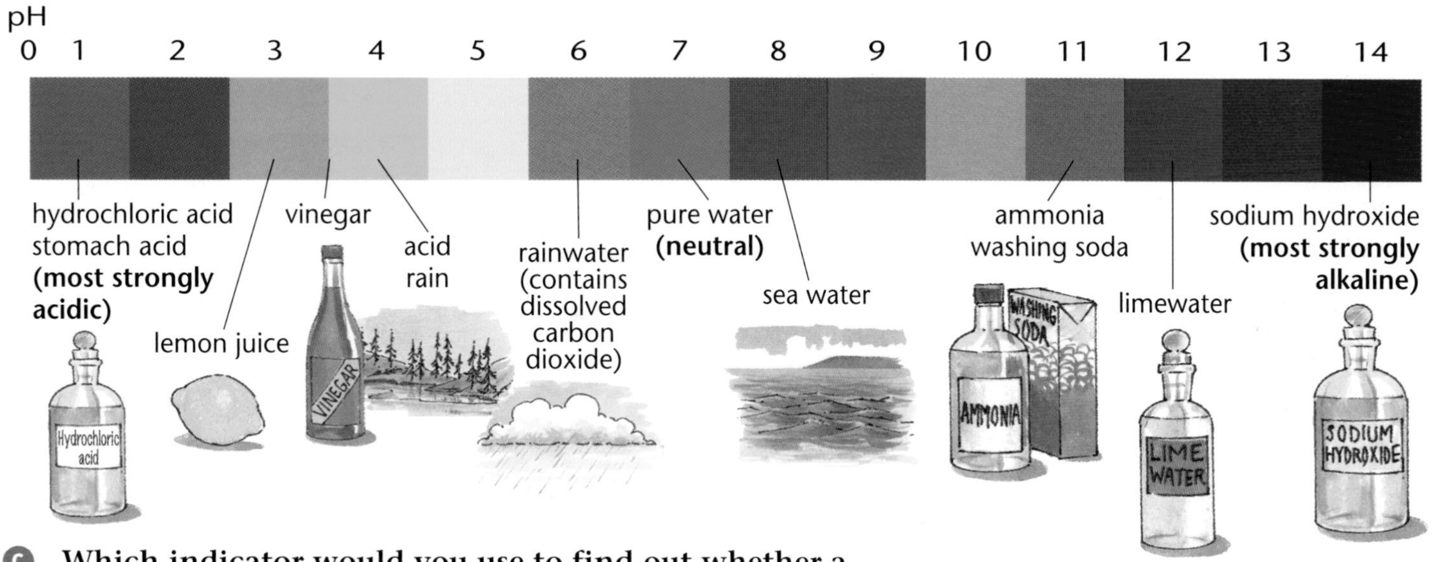

pH

| 0 | 1 | 2 | 3 | 4 | 5 | 6 | 7 | 8 | 9 | 10 | 11 | 12 | 13 | 14 |

hydrochloric acid stomach acid **(most strongly acidic)**

lemon juice

vinegar

acid rain

rainwater (contains dissolved carbon dioxide)

pure water **(neutral)**

sea water

ammonia washing soda

limewater

sodium hydroxide **(most strongly alkaline)**

c Which indicator would you use to find out whether a solution is strongly or weakly acidic?

d What does the pH number tell us?

e Which word describes a substance with pH 7?

Questions

1 Look very carefully at the pH chart above. Copy the table below and complete it. The first line has been done for you.

Liquid	Colour with universal indicator	pH	Description
indigestion medicine		9	weakly alkaline
tea			
salt water			
rainwater			
soap			
lemon juice			

2 Universal indicator paper turns blue with floor cleaner. What is the pH of floor cleaner?

3 Apple juice has pH 3. What colour will universal indicator turn with apple juice?

For your notes:

- An acidic solution may be strongly or weakly acidic. An alkaline solution may be strongly or weakly alkaline.

- **Universal indicator** turns different colours with different strengths of acidity and alkalinity.

- The **pH scale** is used to measure the strengths of acidic and alkaline solutions.

- Water is neutral, so it has pH 7.

E3 Taking away acidity

Putting acids and bases together

We say that a base is the opposite of an acid, because if you add a base to an acid you take away its acidity and make a new substance.

a **What do you think happens if you mix a base with vinegar?**

Look at the photo. If you add vinegar to the base sodium hydrogencarbonate, you will see bubbles. The vinegar, which is an acid, is used up. The pH falls and you get a neutral solution. This is called **neutralisation**.

b **Find the word that describes what happens when you mix an acid and a base.**

Acid indigestion

Your stomach makes hydrochloric acid, which helps you digest your food. The stomach has a special layer which stops the acid corroding your insides!

Sometimes the stomach produces too much acid and you might get acid indigestion. This happens because you have eaten too much, not because you have eaten acidic food.

c **What causes acid indigestion?**

Indigestion medicines are sometimes called 'antacids' or 'anti-acids'. They have alkalis or bases in them to neutralise the stomach acid.

d **How do we treat acid indigestion?**

> **Do you remember?**
>
> When you mix materials, this often causes them to change.

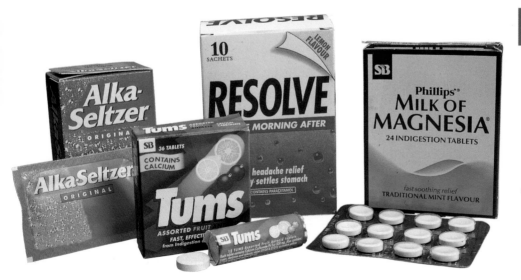

> **Did you know?**
>
> Before all these medicines, people just used to take a spoonful of sodium hydrogencarbonate in a glass of water. It was cheap and it worked!

Problem soil

Some soils are slightly acidic. This is fine for plants such as heathers, which grow in places like the Yorkshire moors. The rhododendrons in this photo also like acidic soil. But most plants and food crops will not grow in acidic soil.

e **Why can an acidic soil be a problem for a gardener?**

f **What would you do to neutralise an acidic soil?**

Farmers and gardeners dig in a base called **lime** to neutralise acidic soil.

Waste water treatment

Water from factories is often acidic or alkaline. One way of neutralising acidic waste water is to pass it over a bed of limestone, as shown below.

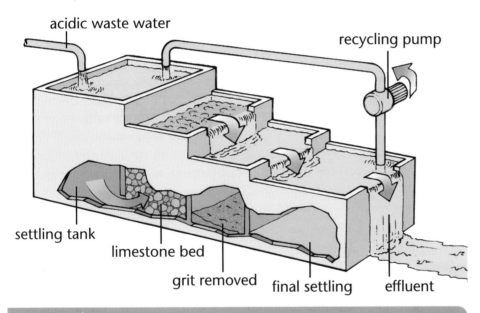

acidic waste water

recycling pump

settling tank

limestone bed

grit removed

final settling

effluent

Questions

1 Copy and complete these sentences by choosing from the words below.

> **acid   alkali   base   neutralisation**

A _____ is the opposite of an _____. A _____ will react with an _____, taking away its acidity. This change is called _____.

2 Val had acid indigestion. Roger told her not to drink cola. Why do you think he did this? What should she do?

3 Why doesn't stomach acid corrode the inside of your body?

4 What could you do to neutralise alkaline water from a factory?

For your notes:

- When you add a base to an acid, a change called **neutralisation** takes place.

E4 Problem soil

Soil acidity

James Beck is worried about the acidic soil on his farm. He has decided to use 'Superbase' to neutralise the soil. He wants to use enough base to neutralise all the acid in the soil, but not too much because that would make the soil alkaline. It would also waste money.

Ask an expert

James has asked soil scientist Sarah Jones to advise him.

Sarah tells him that to neutralise the acid particles, James must add the same number of 'Superbase' particles. The diagram opposite shows this.

acid particles in the soil

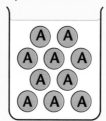

same number of base particles needed to neutralise acid particles

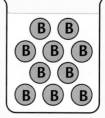

Sarah calculates that one bag of 'Superbase' dissolved in 10 litres of water will neutralise the soil in the field.

> I will mix a **sample** of soil with water, and then measure the number of acid particles dissolved in the water. Then I can estimate the amount of base we need.

ⓐ **Sarah did her calculation based on a sample of the soil in the field. Do you think her estimate is reliable?**

James is still keen to save money. He decides to add another 10 litres of water to Sarah's solution. He then has 20 litres of 'Superbase' solution. He uses 10 litres of it now and saves the other 10 litres for next time.

ⓑ **Do you think this will make any difference to the amount of soil he can neutralise?**

False economy

Sarah was surprised when she found that James' soil was still acidic. So James told her what he did.

*If you add twice as much water, you must **compensate** by adding more base. Or you can use twice as much of your solution.*

Sarah explained, 'My solution had the right number of base particles to neutralise the acid particles in the soil. When you added another 10 litres of water, you still had the same number of base particles because you only used one bag of 'Superbase'.

'When you add more water, the number of base particles stays the same. They are just more spread out. The solution is more dilute.'

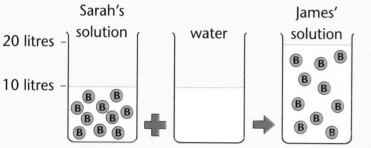

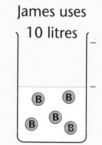

20 litres –

10 litres –

Sarah's solution   water   James' solution   James uses 10 litres

c How much of the soil do you think James' 10 litres of solution would neutralise?

Dilution factors

If James dissolves one sack of 'Superbase' in 10 litres of water and then adds another 10 litres of water, he must compensate and use twice as much solution to neutralise the soil.

The total amount of solution James has is: 10 litres + 10 litres = 20 litres. This is 2 times the solution he had first, so the dilution factor is 2.

d If James added 90 litres of water to Sarah's 10 litres of solution, how much do you think he would need to use to neutralise the acid particles in the soil?

e Can you think of any other situations where you have to compensate for dilution?

Questions

Discuss these questions with your partner and decide which is the correct word from each pair. Write down your answer to describe the idea of **compensation**.

1 If you add more water to a base solution, the base particles will be **more/less** spaced out.

2 You will need to compensate by using **more/less** of the solution to neutralise the soil.

F1 Changing materials

Physical and chemical changes

When you boil water a **physical change** happens. You turn water into water vapour. You can then turn the water vapour back into water.

It is a **reversible change**. No new substances were made.

Burning a match is a **chemical change**. During a chemical change you make new substances. The wood has gone. It has changed into ash, smoke and gases.

It is an **irreversible change**. You cannot turn ash back into wood.

a Divide these changes into physical changes and chemical changes.

> *freezing ice   cooking a burger*
> *burning charcoal   melting an ice lolly*

Chemical reactions

Every time a new substance is made, there is a **chemical reaction**. A chemical change, like burning a match, may involve lots of chemical reactions.

b Copy and complete this sentence: If a new substance is made, a ____ ____ has happened.

Spotting a chemical reaction

Rusting is a chemical reaction. Look at the photo of the old car. The new substance made is rust. You can see the rust because it is orange.

Look at the picture on this packet. It shows a chemical reaction. A new substance is made, which is a gas. This gas makes bubbles.

Colour changes and bubbles are often signs of a chemical reaction.

When a firework explodes, there is a chemical reaction. New substances are made, but they are high up in the air and you cannot see them.

You can see the coloured light given out during the chemical reaction. Energy being given out is often a sign of a chemical reaction.

c Give three signs of a chemical reaction.

Describing chemical reactions

Chemical reactions make new substances. These new substances are called **products**. The substances you started with are called **reactants**. Reactants are used up during the reaction.

d What do we call (i) the substances used up in a reaction and (ii) the substances produced in a reaction?

Questions

1 Copy and complete these sentences using the words below.

   light   chemical   used up   heat   irreversible   new

   Burning a match is a _____ reaction. It is an _____ change. The wood is _____ _____ and _____ substances are made. Energy is given out as _____ and _____.

2 Look at the photo on the right. It shows a sparkler burning. How do you know this is a chemical reaction?

For your notes:

- A **chemical reaction** is a change that makes a new substance.

- Colour changes and bubbles are signs of a chemical reaction.

- Energy being given out is also a sign of a chemical reaction.

- In a chemical reaction, the substances that are used up are **reactants** and the new substances are **products**.

F2 Acids and metals

Learn about:
- Acids reacting with metals
- A test for hydrogen
- Corrosion

Hubble, bubble!

The photo on the right shows zinc being added to sulphuric acid. The zinc is used up and bubbles form.

(a) How can you tell that there is a chemical reaction between zinc and sulphuric acid?

The bubbles contain gas. You can collect the gas by passing it through a tube, as shown in the photo below.

You have to start with the upside-down test tube full of water. The gas pushes the water out of the test tube.

zinc + sulphuric acid

water

hydrogen

The gas is called **hydrogen**. There is another product of this reaction. It is called zinc sulphate. You cannot see the zinc sulphate because it is colourless and dissolves in the water.

(b) Name two reactants in this chemical reaction.

(c) Name two products of this chemical reaction.

Other acids and metals

Other acids also react with other metals to make hydrogen. For example, hydrochloric acid reacts with zinc to make hydrogen and zinc chloride. Hydrochloric acid reacts with aluminium to make hydrogen and aluminium chloride.

c **What product is made in all three reactions between an acid and a metal?**

A test for hydrogen

If you put a lighted splint near the top of the test tube you will hear a 'pop'! Hydrogen is the only gas that pops like this.

d **How do you use a lighted splint to test for hydrogen gas?**

This test for hydrogen is another chemical reaction. You can tell because energy is given out – that's the 'pop' sound.

Corrosion

The sulphuric acid seemed to 'eat away' the zinc during the chemical reaction. The zinc was used up as the hydrogen was made. When metals are used up by chemical reactions we call it **corrosion**.

Look at the photo on the right. It shows a metal tray that a car battery sits on. The tray has been corroded by acid escaping from the car battery.

Questions

1 Copy and complete these sentences using the words below.

pop chemical hydrogen acids corroded

Some metals react with _____ to make hydrogen.

We test for _____ using a lighted splint. A _____ means it is hydrogen.

We say that the acid has _____ the metal, because the metal is used up in the _____ reaction.

2 Describe how you would test a gas to see if it was hydrogen.

For your notes:

- Some metals react with acids to make **hydrogen** gas.

- Hydrogen gas pops when you test it with a lighted splint.

- Metals **corrode** because of chemical reactions.

Learn about:
● Acids reacting with carbonates
● A test for carbon dioxide
● Carbonates

Fizz!

Look at the photo. It shows a piece of chalk in hydrochloric acid. There are bubbles of gas and the chalk is used up.

ⓐ **How do you know that there is a chemical reaction between the chalk and the acid?**

The gas is **carbon dioxide**. Like hydrogen, carbon dioxide is a colourless gas. Unlike hydrogen, carbon dioxide does not burn. If you put a lighted splint into a test tube of carbon dioxide, it will go out.

ⓑ **What happens when you put a lighted splint into: (i) hydrogen? (ii) carbon dioxide?**

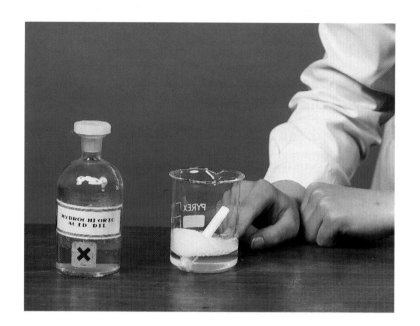

A test for carbon dioxide

You test for carbon dioxide by bubbling the gas through **limewater**. Look at the photo on the right. Carbon dioxide makes limewater go cloudy.

Carbonates

Chalk is calcium carbonate. There are lots of **carbonates**, for example, copper carbonate, sodium carbonate and sodium hydrogencarbonate. All carbonates make carbon dioxide when they react with acids.

When a carbonate reacts with acid, both the carbonate and the acid are used up. The acid goes away, so it is a **neutralisation** reaction.

Light and fluffy

We use baking powder to make cakes rise. Baking powder contains an acid and a carbonate. The acid and the carbonate react together when we add the baking powder to the cake mixture. Carbon dioxide is made. The bubbles of carbon dioxide make the cake rise.

c What type of reaction happens between the two parts of baking powder?

d What product of the reaction makes the cake rise?

Carbonate rocks

Limestone, marble and chalk are made from calcium carbonate. These rocks react with acids in the surroundings because they are carbonates.

The photo shows the Sphinx, a huge statue that was made thousands of years ago. The calcium carbonate is slowly being used up in chemical reactions.

Questions

1 Calcium carbonate reacts with hydrochloric acid to make carbon dioxide.

 a What is the name of the carbonate in this chemical reaction?

 b Name two reactants in this chemical reaction.

 c Name one product of this reaction.

2 Describe the test for carbon dioxide.

3 Explain why statues made from marble wear away over time.

For your notes:

- Acids react with **carbonates** to make **carbon dioxide**. This is a **neutralisation** reaction.

- Carbon dioxide gas turns **limewater** milky.

- Limestone, marble and chalk are made from calcium carbonate, so they react with acids. This means that they wear away because of acids in the surroundings.

Oxygen supply

Is air needed for burning? Look at the picture. Three jars of different sizes are placed over a burning candle.

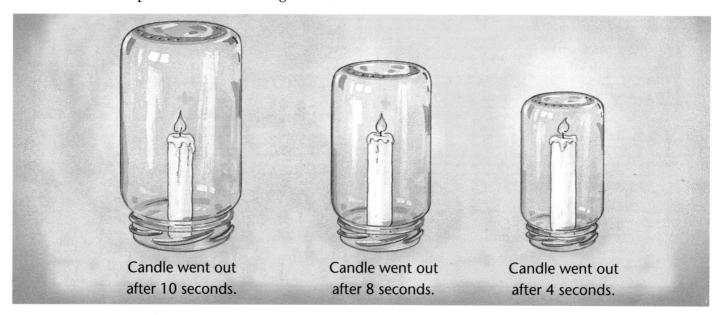

Candle went out after 10 seconds.   Candle went out after 8 seconds.   Candle went out after 4 seconds.

ⓐ Which candle burned for the longest time?

The candle under the biggest jar burned for the longest time. This jar has the most air in it. The candle needs air to burn.

Burning uses up part of the air. Air is a mixture of gases you cannot see. One of these gases is used up during burning. We need the same gas to stay alive.

ⓑ What is the gas in the air that we need to stay alive?

The gas in the air that is used up during burning is **oxygen**.

ⓒ Look at the pie chart. How much of the air is oxygen?

If the jar was filled with pure oxygen, the candle would burn brighter and stay alight longer.

Combustion

Burning is a chemical reaction. The reactants are the material that burns, called the **fuel**, and oxygen. The fuel and the oxygen are used up during the chemical reaction.

The scientific name for burning is **combustion**.

ⓓ Name one of the reactants in the chemical reaction called combustion.

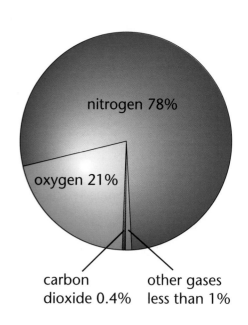

nitrogen 78%

oxygen 21%

carbon dioxide 0.4%   other gases less than 1%

The fire triangle

Fires need fuel and oxygen. But fuel and oxygen are not enough to get a fire going. You need to 'light' a fire to get it going. The chemical reaction needs a little bit of energy (a spark) to get it going. We show this as a **fire triangle**.

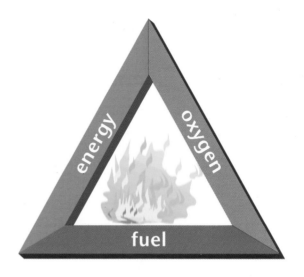

Fighting fires

The fire triangle shows that there are three ways of putting out a fire:

- take away the fuel

- take away the oxygen

- cool it down so there is not enough energy to start the next bit of fuel burning.

Carbon dioxide fire extinguishers fill the room with carbon dioxide. The carbon dioxide replaces the air.

e How does a carbon dioxide fire extinguisher put out a fire?

Questions

1 Copy and complete these sentences using the words below.

   fuel   spark   chemical   oxygen   combustion

   Burning is a _____ reaction, also called _____. _____ and _____ are used up during burning. To get a fire going you also need a _____.

2 Look at the photo of the firefighter. He is carrying a tank of oxygen on his back. Why does he need oxygen to breathe inside the burning building?

3 The fire triangle tells us that oxygen, fuel and energy are needed for a fire. **a** to **c** are three ways of putting out a fire. Explain how they work.

   a Digging a 'firebreak' (removing all the trees and buildings).

   b Soaking with water.

   c Covering with a fire blanket.

For your notes:

- Burning is a chemical reaction. Another word for burning is **combustion**.

- Combustion uses up **oxygen** and **fuel**.

- For a fire to start it needs fuel, oxygen and a little energy (a spark). You can put out a fire by removing one of these three things.

Products of combustion

What are the products of combustion?
It depends on which fuel you use.

One of the simplest fuels is **charcoal**. We use charcoal on our barbecues, like the one in the photo. Charcoal is almost pure carbon. When you burn carbon you make carbon dioxide.

Describing chemical reactions

We can use a **word equation** to tell people about a chemical reaction.

carbon + oxygen	→	carbon dioxide

We put the substances we start with – the **reactants** – on the left.

We show the chemical reaction by an arrow.

We put the new substances – the **products** – on the right.

ⓐ Look at the word equation above. (i) Name the reactants. (ii) Name the product. (iii) What does the arrow show?

Burning metals

As well as fuels, we can also burn metals. The metal magnesium burns with a bright white flame. The magnesium reacts with oxygen in the air. A white powder is made. This is magnesium **oxide**.

We can use a word equation to describe this reaction.

magnesium + oxygen → magnesium oxide

ⓑ What is the product of this reaction?

Oxides

Carbon dioxide and magnesium oxide are both products of combustion. In fact, all the products of combustion are oxides.

Oxides are made when oxygen reacts with other substances in a burning reaction.

The name oxide comes from the word oxygen. We take the first two letters of oxygen and add 'ide'

ⓒ What do all oxides have in common?

Making predictions

Carl and Emma were burning substances. They were trying to predict the products of combustion.

First they burned copper. The photo shows copper being burned. A black substance forms on the surface.

It will make copper oxide.

d Copy and complete this word equation for copper burning.

copper + oxygen → _____ _____

Then they watched their teacher burning sulphur. Look at the photo below. Sulphur is a yellow substance. It burns with a blue flame.

It will make sulphur oxide.

Oxide or dioxide?

Why are some oxides called 'dioxides' and others called 'oxides'? The answer is that some substances can make two different oxides, so you have to show which one. For example, carbon can burn to make carbon monoxide or carbon dioxide.

We call it sulphur dioxide.

Did you know?

Water is an oxide. It is really hydrogen oxide, but everyone called it water long before scientists worked that out!

Questions

1 Copy and complete these word equations.

magnesium + oxygen → _____ _____

carbon + oxygen → _____ _____

sulphur + _____ → sulphur dioxide

2 Iron burns. It reacts with oxygen in the air. Predict the product of this reaction.

3 Write your own word equations for these three reactions.

  a Zinc reacts with oxygen to make zinc oxide.

  b Calcium reacts with oxygen to make calcium oxide.

  c Hydrogen reacts with oxygen to make water.

For your notes:

- **Oxides** are made during burning. They are the products of combustion.

- We can show chemical reactions using **word equations**.

Fuels in industry

Fuels are substances that we burn to release energy that we can use. We burn large amounts of fuels such as coal, oil and gas. This gives **light energy** and **heat energy**, and energy to make things move.

When coal, oil and gas burn they produce carbon dioxide. The more fuel we burn, the more carbon dioxide we release into the air.

The greenhouse effect

The carbon dioxide in the air has the same effect as the glass in a greenhouse. The glass stops some of the heat energy in the greenhouse escaping, and the plants stay warm.

Carbon dioxide stops some of the heat energy from the Earth escaping, and the Earth stays warm. This is called the **greenhouse effect**.

Scientists think that the greenhouse effect may make the Earth too hot as we produce more and more carbon dioxide.

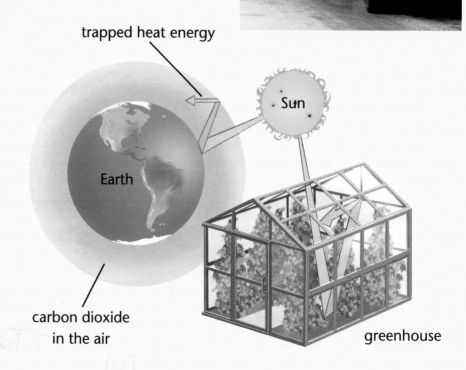

trapped heat energy

Sun

Earth

carbon dioxide in the air

greenhouse

ⓐ What stops some of the heat energy escaping from the Earth?

ⓑ Why do we call this the greenhouse effect?

The layer of carbon dioxide in the air acts like a quilted jacket. The thicker the quilt, the hotter you are!

ⓒ Copy and complete this sentence to show what we think the relationship is between the temperature of the Earth and the carbon dioxide in the air:

As the amount of carbon dioxide increases, the Earth gets ...

ⓓ What do you think would happen if the amount of carbon dioxide in the air decreased?

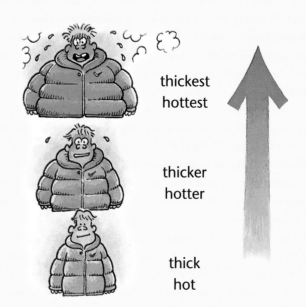

thickest
hottest

thicker
hotter

thick
hot

The biodome experiment

A team of students decided to find out whether the temperature of the Earth rises when the level of carbon dioxide increases.

They set up a biodome. This is a model of the Earth. Carbon dioxide can be added through the tube at the side. The carbon dioxide level and the temperature inside the dome were recorded by a computer.

Any relationship?

e What is the *input variable* in this experiment?

f What is the *outcome variable*?

We use the word **relationship** to describe how the outcome variable changes when we change the input variable.

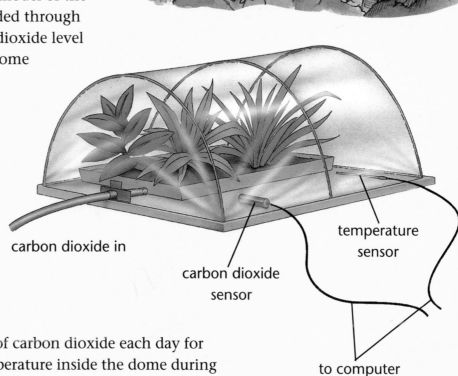

carbon dioxide in

carbon dioxide sensor

temperature sensor

to computer

The students increased the level of carbon dioxide each day for five days. They recorded the temperature inside the dome during that time to see if there was a relationship between carbon dioxide level and temperature. Their results are shown in the table.

Day	Input variable: carbon dioxide level in %	Outcome variable: temperature in °C
1	0.1	20
2	0.2	21
3	0.3	22
4	0.4	23
5	0.5	24

Questions

1 Look at the table above. Copy and complete these sentences by choosing from the words in bold.

   a Both the carbon dioxide level and the temperature in the biodome have **increased/decreased** over the five days.

   b As the carbon dioxide level **increased/decreased**, the temperature became **high/higher**.

2 Find out how the greenhouse effect could change our weather.

67

G1 Developing theories

Different materials

Water, air and this piece of paper are all examples of **materials**. Why can materials behave so differently? Why can water be steam, liquid or ice?

Ideas about materials

About two and a half thousand years ago in Ancient Greece, groups of thinkers, or philosophers, started to think about materials and what they were made of. There were two main ideas.

Idea 1: *The four elements*

THALES · 580BC — Everything is made up of water.

ANAXIMANDER · 520BC — Everything is made from air.

HERACLITUS · 480BC — Everything is made of fire.

EMPEDOCLES · 450BC — Everything is made up of four different substances: fire, air, water and earth.

The idea that everything is made up of fire, air, water and/or earth really caught on. A very famous philosopher called Aristotle believed it. For the next 2000 years students were taught that Aristotle was right.

a **Who came up with idea of 'four elemental substances'?**

Aristotle.

Idea 2: *Atoms*

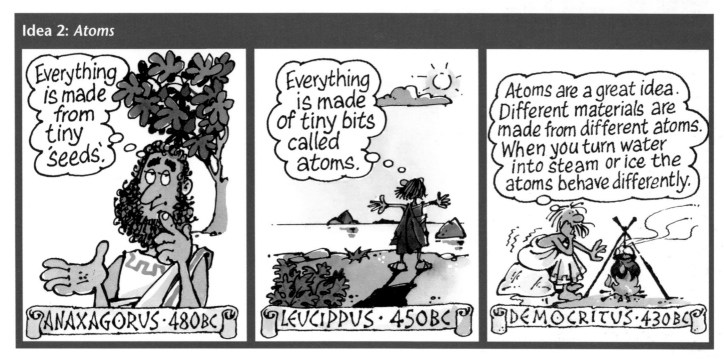

The idea of **atoms** was unpopular for the next 2000 years. People still believed that everything was made of four elements.

ⓑ Who came up with the idea of atoms?

Ideas today

In AD 1803 the scientist John Dalton published his 'Atomic theory of matter' which said that all matter was made of atoms. A **theory** is an idea that explains something. Dalton realised that the way materials behave could only be explained by atoms.

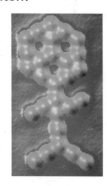

John Dalton.

Today atoms are no longer just a theory. The picture shows atoms seen using a powerful microscope. The scientist has even managed to arrange the atoms to make a stick figure!

Atoms and particles

Although all substances are made of atoms, those atoms can be arranged in different ways. They can be single atoms, or groups of atoms called **molecules**. We are going to use the word **particle** because that covers both atoms and molecules.

Questions

1 Make a time line showing how the idea of atoms developed. Use the information on these two pages.

2 Leucippus and Empedocles lived at the same time, although in separate parts of Europe. Write a story about the two of them meeting. Include what you think they would say to each other.

For your notes:

● Ideas about materials have changed over time.

● All matter is made of **particles**.

● 'Particle' can mean an **atom** or a group of atoms.

Particle power

A closer look

The theory that particles make up all materials is called the **particle model**. The particle model explains why solids, liquids and gases behave in different ways. To find out why, let's have a closer look.

Solids

In a solid, the particles are packed closely together. They are in neat rows and do not move much. The particles are joined together strongly.

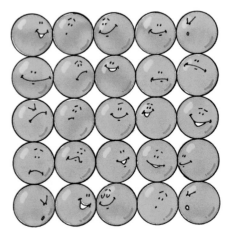

a Describe how the particles are arranged in a solid.

Liquids

In a liquid, the particles are close together. They are still touching. The particles are joined together less strongly than in a solid. The particles slide over each other. They are not in a regular pattern.

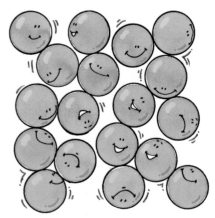

b Describe the movement of the particles in a liquid.

c Compare how the particles are held together in a solid and a liquid.

Gases

In a gas, there is nothing holding the particles together. The particles are moving quickly. The particles are very far apart and not in any pattern.

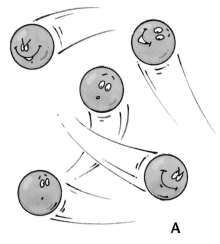

d Describe the movement of particles in a gas.

e Is there anything holding the particles together in a gas?

A

Density

Look at the pictures. One cube has lots of particles packed into a small volume. This makes it heavy. We say that the cube is **dense**.

The other cube is the same size but there are only a few particles inside. It has a smaller **density**. This makes it lighter than the other cube.

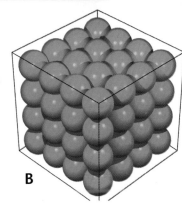

B

f Which cube shows a solid?

Gases are less dense than solids or liquids.

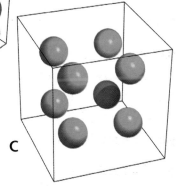

C

Questions

1 Copy and complete these sentences by choosing from the words in bold.

   In **solids/gases** the particles are touching.

   In **liquid/gases** the particles are separated.

   The particles in a **solid/liquid/gas** are held together most strongly.

   In **liquid/solids** the particles are in neat, ordered rows.

   In **solids/liquids/gases** the particles are moving most quickly.

2 Jack says, 'The particles are sliding over each other. They are not ordered. They are held together weakly.'

   Is Jack describing a solid, a liquid or a gas?

3 **a** Which is heavier, 1 kg of water or 1 kg of steam?

   b Which takes up the most space?

   c Which is more dense?

For your notes:

- In a **solid**: the particles stay in their places, are joined together strongly and are closely packed in a pattern.

- In a **liquid**: the particles slide over each other, are joined together weakly and are close together but not in a pattern.

- In a **gas**: the particles are far apart, moving quickly and not joined together.

- Solids and liquids are more **dense** than gases.

G3 Looking at evidence

Solids

How does the particle model fit with what we know of solids?

a Why are solids strong?   **b** Why don't solids squash?

Solids don't squash.

Solids are strong.

Solids keep their shape and volume.

That's because the particles are joined together by strong pulling forces.

That's because the particles are touching and in a regular pattern. Every particle stays in its place because it's joined to the next one.

That's because the particles are already touching. You can't push them closer together.

Liquids

How does the particle model fit with what we know about liquids?

c Why do liquids change shape?   **d** Why do liquids stay the same volume?

I can stir a liquid but not a solid.

That's because the particles in a liquid aren't joined together as strongly as in a solid.

Liquids keep their volume but not their shape.

It won't squash. It feels like a solid.

They keep their volume because the particles are held closely together. They change shape because the particles can slide over each other.

That's because the particles are already touching. You can't push them closer together.

Gases

How does the particle model fit with what we know about gases?

e **Why can you squash a gas?**

f **Why can we walk easily through the air?**

It feels like there's nothing there.

There is something there. I can feel my breath.

This balloon squashes.

That's because the particles aren't held together. They just move away from your hands.

That's because the particles are moving quickly. They are hitting your hand.

That's because there are spaces between the particles. You can push them closer together.

Using the model correctly

We can use the particle model to explain other observations, such as expansion.

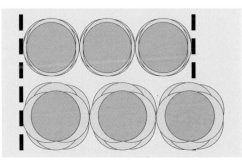

The hotter particles vibrate more and take up more space.

*Solids **expand** when they are heated. A zinc bar 1 m long at 20 °C is 3 mm longer when it is heated to 120 °C.*

Yes, they just vibrate more so they take up more space.

So the particles stay the same size and are still touching?

That's easy – the particles get further apart.

That's not right. Don't forget that particles are always touching in a solid. If there were gaps between them, it would be a gas!

Questions

1 Look at the cartoon. The man is going to hurt himself because the water will not squash. Why doesn't the water squash?

2 Look at this photo of a ship. The wind pushes the sails, making the boat move through the water. Why does the wind push the sails?

For your notes:

- We can use the particle model to explain how solids, liquids and gases behave.

- You have to be careful to use the particle model correctly.

- Solids **expand** because the particles vibrate more when they are heated.

What's that smell?

All the things in these photos give off smells. You cannot see what is making the smell in the air. The smell is caused by tiny particles that are too small to see. Special cells in your nose can detect these particles.

Imagine someone lets off a stink bomb. At first, only the people close-by would smell it. Later, everyone would smell it.

Look at the diagrams below. They show how the smelly gas particles mix with particles in the air. Particles of a gas move very quickly in all directions. They reach every part of the room.

This spreading and mixing is called **diffusion**. Diffusion happens by itself – you don't need to mix or stir the substances.

a Why would a person sitting near the stink bomb smell it before someone sitting further away?

Diffusion in liquids

These photos show diffusion happening in a liquid. The purple substance slowly **diffuses**, or spreads out. This is because the particles in a liquid can move and change places.

b Why is diffusion faster in a gas than in the liquid?

c Why was the purple colour at the top of the beaker stronger at the end than it was at the beginning?

Pump it up

Inside this balloon are millions of gas particles. The gas particles are moving around in all directions. They hit the sides of the balloon.

Each hit gives the side of the balloon a tiny push. Each push is quite small, but lots of particles all pushing at the same time adds up to quite a lot of force. We call this force **gas pressure**. There is gas pressure inside the balloon.

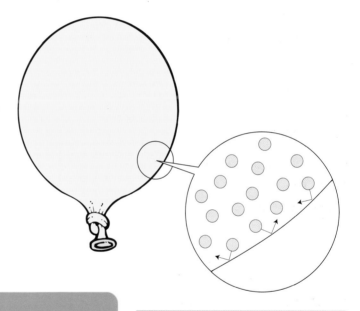

d How do particles move in a gas?

e What causes gas pressure?

Questions

1 Copy and complete these sentences using the words below. Use one of the words twice.

> push   directions   diffusion   particles

Gas _____ can move and spread out in all _____. Eventually, they spread out evenly in the air. This mixing is called _____.

Gas pressure is caused by the _____ of a gas hitting the sides of the container. Each hit gives the container a little _____.

2 Why does diffusion not happen in a solid?

3 Explain why a balloon gets bigger if you put more air in it.

For your notes:

- **Diffusion** is the spreading and mixing caused by the movement of particles.
- Diffusion happens in gases and liquids.
- Gases cause **gas pressure**.
- Gas pressure is because of the gas particles hitting a surface.

G5 Scientific models

Models

The word 'model' can mean lots of different things to us. You might think of a fashion model on a catwalk, or a model aeroplane. When scientists talk about a **model**, they usually mean a way of showing how something works or looks that they cannot see or touch.

Christmas presents

On Christmas Eve, Jackie's six-year-old brother Matt had gone to bed. Jackie was looking at the presents for him under the Christmas tree. She couldn't open them, but she wanted to try and find out what was in them.

First she just looked at three presents for Matt and thought about their shapes. She tried to decide what they might contain.

a **Look at the shapes of the three presents for Matt. What could they contain?**

Next she picked up the three presents and felt how heavy they were. After that she squeezed them.

b **What else could Jackie do to collect information about the presents? (Remember, she can't open them!)**

Eventually, Jackie decided that one of the presents could be a book or a video.

c **In what ways are a book and a video:**
   (i) similar?
   (ii) different?

Finally Jackie decided the present was a video. She had collected her evidence and used her model of a video to explain what she could see and feel about the present.

Scientific models

Scientists use models to help them think about how things work or behave. First they think about the evidence they have collected, and then they use a model that might explain it. You have met the particle model in this unit. Now try using it to explain the following two events.

① Feeling the cold

A A balloon was inflated with air.

B The balloon is dunked into liquid nitrogen. This is very cold indeed.

> *The air particles inside are moving quickly and hitting the sides of the balloon.*

C The cold balloon is a lot smaller.

> *As the particles get colder they begin to move a lot slower. They get much closer together.*

ⓓ The balloon has got smaller. Write a sentence to explain what happened to the balloon in photo C. (*Hint:* remember to use the particle model!)

ⓔ Predict what will happen to the balloon when it warms up. Explain your prediction using the particle model.

② Glass-blowing

The person in the photo on the right is making a bottle. This is called glass-blowing. A ball of glass is heated until it is red-hot. When it is hot, the glass can be stretched or shaped. Once it is the right shape, the glass is left to cool.

ⓕ Is the hot glass like a liquid or a solid? Explain why you think this.

ⓖ The glass changes shape when it is hot, but keeps it shape when it is cool. Use the particle model to explain why.

Questions

1 Scientists use the particle model to make predictions. Predict what will happen when:

  a propanone (nail varnish remover, a liquid) is heated

  b mercury (a liquid) is cooled

  c nitrogen (a gas) is cooled until it is very, very cold.

2 Scientists use the particle model to explain things. Explain the following.

  a You can smell food cooking from the other end of the house.

  b It is difficult to get a metal lid off a jam jar when it has been stored in a fridge.

H1 Pure salt

Pass the salt

We put salt on our food. Most table salt is made from 'rock salt'. Rock salt is dug out of the ground. It is a dirty, gritty mixture. You wouldn't want to put it on your chips!

Rock salt is a mixture of different substances. One of those substances is what we call 'salt'. The salt is separated from the other substances before we use it. We say that the salt is 'purified'. Salt on its own is a **pure** substance.

ⓐ Why isn't the rock salt pure?

Purifying salt

Joe and Catherine are getting salt from rock salt.

This isn't just salt. There are bits of grit and sand in here.

Salt dissolves in warm water. Grit and sand don't.

Then we could filter it. The salt solution will come through.

Then we can evaporate the water to get the salt.

Do you remember?

When a solid dissolves, you can separate it from the solution by evaporating.

When a solid does not dissolve, you can separate it from the solution by filtering.

Joe and Catherine add warm water to the rock salt and stir.

ⓑ Why do Joe and Catherine add water to the rock salt?

ⓒ Look at the photo above of the rock salt in water. How can you tell that it is a mixture?

Then they filter the mixture. The solid stays in the filter paper. The liquid that drips through is clear and colourless. It is salt **solution**. The salt has dissolved in the water because it is **soluble**. The sand stayed in the filter because it was **insoluble**.

ⓓ Why do Joe and Catherine filter the mixture?

Joe and Catherine pour the salt solution into an evaporating dish. They heat the solution. Most of the water **evaporates**. They leave the solution to cool. The next day there are crystals of salt in the evaporating dish.

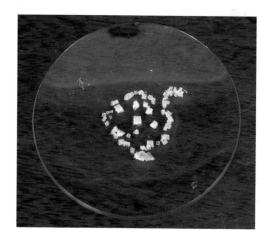

e How did Joe and Catherine separate the water and the salt?

Making a solution

The liquid that does the dissolving is called the **solvent**. Our solvent was water.

The solid that dissolves is called the **solute**. Our solute was the salt.

Reporting back

Joe is writing about the experiment.
He wants to use the correct scientific words.

f Read Joe's notebook. What do you call:
  (i) the solid that dissolves.
  (ii) the liquid that dissolves the solid.
  (iii) the mixture that is made.

Questions

1 Match each word with its correct meaning.

Word	Meaning
evaporating	does not dissolve in a **solvent**
filtering	a substance on its own, with no other substances mixed in
insoluble	dissolves in a solvent to make a clear solution
mixture	heating so that the solvent turns into gas, leaving the **solute**
pure	more than one substance
soluble	passing through paper to separate soluble and insoluble substances

2 Copy and complete this flow diagram using the words below to explain how salt is made from rock salt.

evaporate   dissolve   filter

rock salt —1→ salt solution and undissolved substances —2→ salt solution —3→ salt

For your notes:

- A mixture of **insoluble** and **soluble** substances can be separated by dissolving and filtering.

- A soluble substance, or **solute**, dissolves in a **solvent** to make a **solution**.

- The solvent and the solute can be separated by evaporation of the solvent.

H2 Distillation

Getting drinking water

Imagine that you live on an island. The supply of fresh water has been cut off. There is no fresh water for you to drink. Sea water is not good to drink. It is a solution of salt in water. How could you separate the water and the salt to get drinking water?

If you heat the water to evaporate it, it will turn into a gas (water vapour) and escape. The salt would be left behind. But you want the water. To get drinking water, you can use a process called **distillation**.

Read how this happens in the diagram.

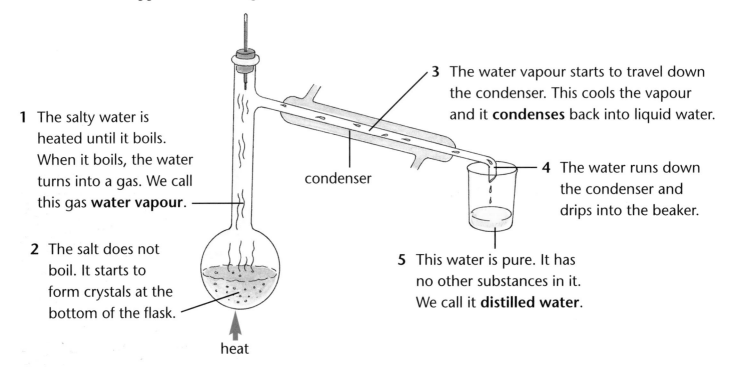

1 The salty water is heated until it boils. When it boils, the water turns into a gas. We call this gas **water vapour**.

2 The salt does not boil. It starts to form crystals at the bottom of the flask.

condenser

heat

3 The water vapour starts to travel down the condenser. This cools the vapour and it **condenses** back into liquid water.

4 The water runs down the condenser and drips into the beaker.

5 This water is pure. It has no other substances in it. We call it **distilled water**.

(a) **What two substances are separated by the distillation?**

Distilled water is pure. It has no other substances mixed in with it. If you boiled pure water, nothing would be left behind.

In some countries there is very little fresh water. Drinking water is made from sea water using distillation. This photo shows a desalination plant where sea water is distilled, to separate the salt from the water.

Separating ink

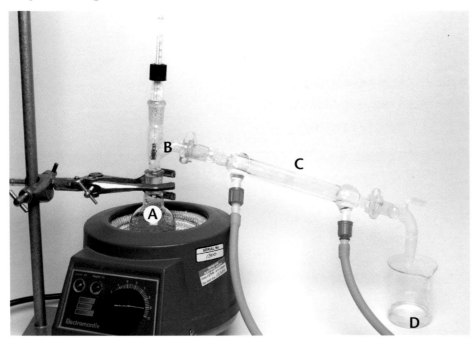

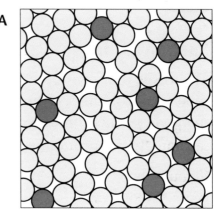

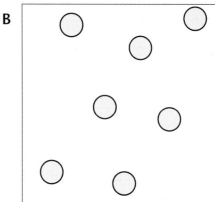

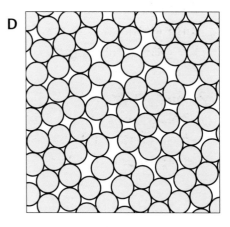

Red ink is a mixture of a red solid dissolved in ethanol. We can separate the mixture by distillation.

b Where in the photo is evaporation taking place?

c Where in the photo is *condensation* taking place?

The diagrams show the particles at **A**, **B** and **D** in the photo.

d Which diagram shows a liquid?

e Which diagram shows a gas?

f Which diagram shows a solution?

g Which diagram shows pure ethanol?

When all the ethanol has been evaporated, a red solid is left at **A**.

h What two substances have been separated?

Questions

1 Copy and complete these sentences using the words below.

gas   vapour   distillation   distilled   condenses

To separate salt and water, we can use a method called _____.
The water is boiled and turns into a _____. The
water _____ reaches the condenser where it cools down
and _____. This pure water is called _____ water.

2 Describe how drinking water can be made from sea water.

For your notes:

- Substances can be separated by **distillation**.

- In distillation a liquid evaporates into a gas and **condenses** back to a liquid.

- Distillation can be used to separate a pure liquid from a solution.

Separating inks

The ink in your pen is probably not made of one colour. It is a mixture of colours or **dyes**. To separate them out, we can use a method called **chromatography**.

1 The mixture is put on the filter paper. The pencil line shows where the mixture was put at first.

2 The solvent travels up the paper. The different coloured inks travel at different speeds up the paper.

Did you know?

Scientists have improved chromatography so it can be used to separate hundreds of different mixtures, not just inks and dyes.

On wet paper, each colour will move a different distance. Colours that are very soluble move a long way. Colours that are not very soluble do not move very far.

Analysing inks

Look at this picture. It shows the separation of four different inks, **A**, **B**, **C** and **D**. The black line and the crosses show where the ink mixtures started.

a Which one of these inks is a mixture?

b How many different dyes is it made of?

c Which dye is most soluble, purple, blue or green?

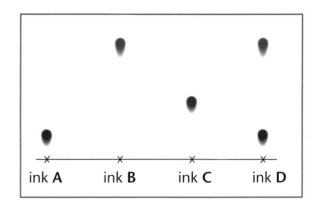

Analysing foods

The Pluto Sweetie company brought out a new range of sweets called 'Brighties'. Tony was worried that the sweets might contain a colouring called sunburst yellow. Tony is allergic to sunburst yellow – it brings him out in big red spots. He can eat other yellow colours.

Tony used chromatography to separate out the colours in the orange and yellow sweets. He compared these with four different yellow food colours. His results are shown below. The four food colours are called:

- sunburst yellow
- solar yellow
- mellow yellow
- sunny yellow.

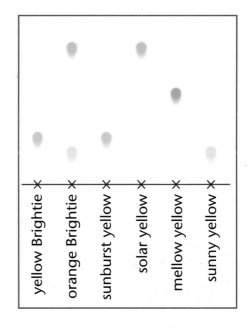

(d) **What colours are in an orange Brightie?** (*Hint:* look across to see which colour has spots at the same height as the orange Brightie.)

(e) **What colours are in a yellow Brightie?**

(f) **Will orange Brighties make Tony ill? Explain your answer.**

(g) **Will yellow Brighties make Tony ill? Explain your answer.**

Questions

1 Put these statements in the correct order to describe a chromatography experiment.

   A Make ink spots on the crosses.

   B Put into a beaker with a small amount of solvent in the bottom.

   C Take out the filter paper when the solvent reaches the top. Let it dry.

   D Take a piece of filter paper.

   E Allow the solvent to run up the paper and separate the dyes.

   F Draw a pencil line at the bottom with small crosses.

2 Is the ink in your pen a mixture? Plan an experiment to find out.

For your notes:

- **Chromatography** is used to separate a mixture of **dyes**.

- The dyes are separated because they are more or less soluble in the solvent.

- Chromatography can be used to tell you which dyes are present in a mixture of colours.

Changing cheques?

Shaheen is a forensic scientist. It is her job to help the police solve crimes. Shaheen's latest case is about a cheque for £1999.

Mr Jones says that he wrote the cheque for £1000. Mr Crisp says that the cheque was always for £1999.

If Mr Jones is telling the truth, then the words 'nine hundred and ninety nine' were added later by Mr Crisp.

If Mr Crisp is telling the truth, then all parts of the cheque were written at the same time.

One ink or two?

If two different people wrote on the cheque, they probably used different pens with different inks. Shaheen decides to analyse the ink using chromatography.

First, she photographs the cheque. Then, she cuts out the words 'one thousand' and the words 'ninety nine'. She wants to see if the ink in 'one thousand' and the ink in 'ninety nine' are different. Shaheen separates the inks from the paper.

a Write out these steps in the correct order to describe how Shaheen separated the ink from the paper.
- Shaheen filtered out the lumps of paper.
- Shaheen cut the words out of the cheque.
- Shaheen dissolved the ink in a solvent.

Shaheen uses chromatography to separate the dyes in the two inks. She uses the solvent propanone.

b Describe how you would separate the inks if you were doing this experiment at school. Write down the steps you would take.

Under normal light, the pattern for ink **A** and ink **B** look the same. Shaheen also looks at the chromatography paper under UV light.

The page from her laboratory record is shown below.

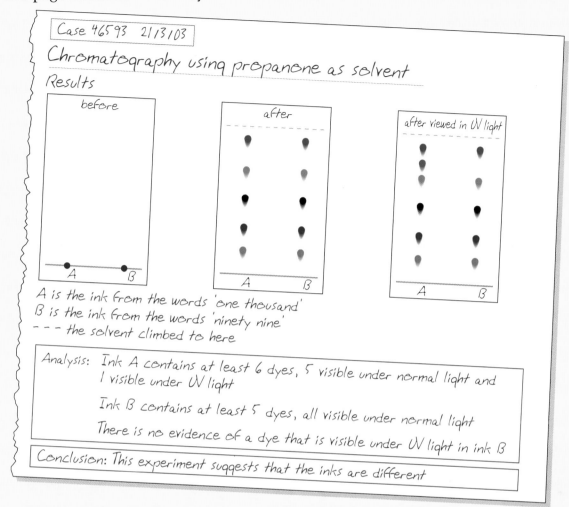

Case 46593   2/13/03

Chromatography using propanone as solvent

Results

before

after

after viewed in UV light

A is the ink from the words 'one thousand'
B is the ink from the words 'ninety nine'
– – – the solvent climbed to here

Analysis: Ink A contains at least 6 dyes, 5 visible under normal light and
1 visible under UV light

Ink B contains at least 5 dyes, all visible under normal light

There is no evidence of a dye that is visible under UV light in ink B

Conclusion: This experiment suggests that the inks are different

c How many dyes are there in ink B? Explain your answer.

d How many dyes are there in ink A? Explain your answer.

e <u>Must</u> the extra dye in ink A come from the cheque?
Suggest other places it could have come from.

f How would viewing the cheque using UV light confirm
that ink A and ink B are different?

g Do you think that the words on the cheque were written
by two different people? Explain your answer.

Questions

1 Mr Jones may have changed pens half-way
through writing the cheque.

a Suggest a reason why Mr Jones may have
changed pens.

b Do you think this is likely? How could you
find out?

2 Imagine you are Shaheen when the case goes
to court. What would you say to the jury in
the court? Don't forget to explain the
evidence for your conclusion.

H5 Solubility

Learn about:
- Dissolving
- Solubility

What happens during dissolving?

The photo on the right shows copper sulphate dissolving in water. The diagrams show what is happening to the particles. The particles in the solid spread out. The particles of copper sulphate (dark blue) mix with the particles of water (pale blue).

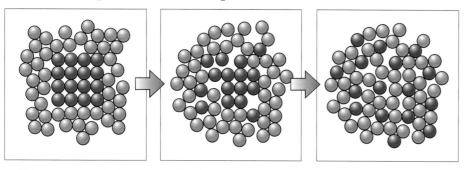

solid copper sulphate   solid breaks up   particles mixed

Where does it go?

Look at the photos below. They show the mass of solid salt and the mass of water, and then the mass of the solution when they are added together.

Solid salt.

Water.

Salt and water.

a What is the mass of:
  (i) the solute?   (ii) the solvent?   (iii) the solution?

Matter is not lost when you make a solution. You have the same mass of solute and solvent as when you started. We say that the mass has been **conserved**.

How much will dissolve?

You can add more and more solute until no more will dissolve. The solution is then **saturated**. Scientists measure the **solubility** of the substance. The solubility is the mass of solute that will dissolve in $100\,cm^3$ of water.

Different temperatures

The graph shows how much of a substance (in grams) dissolves in 100 cm³ of water at different temperatures. The solubility changes with temperature.

Look at the graph.

ⓑ How much of the substance dissolves at 30 °C?

ⓒ How much of the substance dissolves at 70 °C?

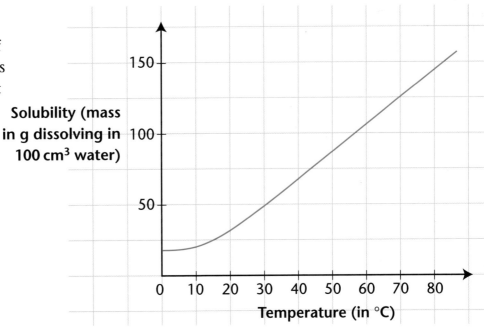

Heating speeds up dissolving. In hotter water the water particles move around more. They hit the solid harder and more often, knocking off particles and breaking up the solid faster.

ⓓ Will sugar dissolve quicker in hot water or cold water?

Different solvents

Different solvents dissolve different solutes. A substance can be soluble in one solvent and insoluble in another.

Common solvents include water, ethanol and propanone. Propanone is often used as nail varnish remover. Nail varnish does not dissolve in water but it does dissolve in propanone.

Questions

1 Copy and complete these sentences using the words below.

  saturated solution solute solvent quickly solubility

  When a substance dissolves, the particles from the _____ are surrounded by particles of _____ to make a _____.

  When the solvent is hotter, the solute dissolves more _____.

  When no more of a substance will dissolve, the solution is _____.

  The amount of a substance that will dissolve is called the _____.

2 If you add 10 g of copper sulphate to 100 g of water, what is the mass of the final solution? Explain your answer.

For your notes:

- **Solubility** is the mass of solute that will dissolve in 100 cm³ of solvent.

- Solubility changes with temperature.

- Solutes dissolve more quickly when the solvent is hotter.

- Mass is always **conserved** during dissolving.

- A substance can be soluble in one solvent and insoluble in another.

I1 Energy on the move

Talking about energy

In science, we use the word 'energy' in a special way. **Energy** makes things happen.

That boy has lots of energy.

You were very energetic today.

Eat your breakfast. It will give you energy.

I haven't the energy to do that!

ⓐ Look at the pictures of the people. Who do you agree with?

On the move

If something is moving, it has energy. We call this **movement energy** or **kinetic energy**. The athlete in the photo is running. He has lots of kinetic energy.

ⓑ Do you have more kinetic energy when you sit or when you dance?

Sounding out

If something makes sound, it is giving out energy. We call this **sound energy**. The hi-fi in the picture gives out sound energy.

ⓒ Which part of your body detects sound?

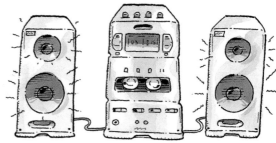

Lighting up

If something is giving out light, then it is giving out energy. We call this **light energy**. The electric lamps in the photo on the left are giving out light energy.

ⓓ Which part of your body senses light?

Hotting up

When something is hot, it is giving out energy. We call this **heat energy** or **thermal energy**.

ⓔ Look at the picture of the iron. How would you be able to tell the iron was working?

Plug it in

Look again at the picture of the iron. It only works if it is plugged in and turned on. It needs energy from electricity to work.

We call this energy **electrical energy**. Electrical energy comes from power stations or from batteries. Look at the photo of the electricity meter. It measures the amount of electrical energy coming into the house from the power station.

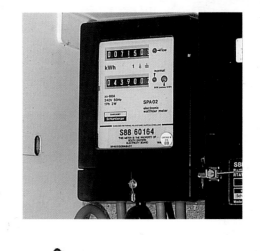

ⓕ   **What other things on these pages need electricity to work?**

Moving energy around

Energy moves from place to place. Look at the drawing of the lamp. The energy comes into the lamp as electrical energy. It comes out of the lamp as light energy and thermal energy. We say that the energy is **transferred**.

We show **energy transfers** using arrows. We write on the arrows to show how the energy is coming in and going out. This is an **energy transfer diagram**.

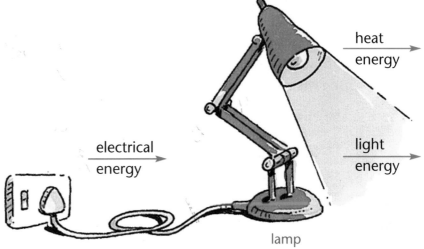

heat energy

electrical energy

light energy

lamp

Questions

1 Look at pictures **A** to **D**. Which picture shows something:

 A    B   C   D

a with lots of kinetic energy?

b giving out light energy?

c giving out sound energy?

d giving out thermal energy?

e working because of electrical energy?

2 Copy and complete this energy transfer diagram.

_____ energy

_____ energy

_____ energy

kinetic energy

kettle

For your notes:

- **Energy** makes things happen. Things work because of energy.

- We sense **light energy, sound energy** and **heat** (thermal) **energy**.

- Things that move have **movement** (kinetic) **energy**.

- Electricity carries energy. We call this **electrical energy**.

- When energy is moved about, we say it is **transferred**.

I2 Stored energy

Taking the strain

Look at the picture of the archer. Where was the energy before it was moving the arrow? It was stored in the stretched bow and bowstring. We call energy that is stored in something stretched **strain energy**.

Look at the picture of a toy frog. It moves because it has been wound up. Winding the toy stores energy in a spring inside the toy.

a **Where is the energy stored in:**
   (i) a bow? (ii) a wind-up toy?

Fuelled up

Look at the picture of a bonfire. The fire is giving out light energy and heat (thermal) energy. Where was the energy before the bonfire started? The energy was stored in the wood. We call energy stored in materials **chemical energy**.

Wood is a **fuel**. All fuels are stores of chemical energy. They give out energy when we burn them.

Food is also a store of chemical energy. Our bodies release the energy stored in the food.

A third store of chemical energy is **batteries**. Batteries use the stored chemical energy to give out electrical energy.

b **Name three different ways of storing chemical energy.**

Lifted up

Look at this picture. The bucket and water have movement energy. Where was the energy before the water and the bucket started moving?

The energy was stored in the bucket and the water because they were lifted up. Things that are lifted up have energy because of **gravitational attraction (gravity)**. We call energy stored because of gravity, **gravitational energy**.

c Which would have more stored energy, a ball on a table or a ball on the floor?

Showing stored energy

We can show stored energy in a box. The diagram below shows the stored energy in a bonfire. It is shown in the box. The energy given out when the wood is burned is shown by the arrows.

chemical energy in wood	heat energy →
	light energy →
	sound energy →

Questions

1 Copy and complete these sentences.

　a Strain energy is …

　b Chemical energy is …

　c Gravitational energy is …

2 How is the energy stored in **a**, **b** and **c**? Choose from gravitational energy, chemical energy or strain energy.

　a a skydiver jumping out of a plane

　b a firework

　c a squashed ball.

3 Copy and complete this energy transfer diagram for a bow and arrow.

```
 _____
|      |
|_____|_____ energy →
|_____|
```

For your notes:

- Energy stored because a material is being pulled or pushed is called **strain energy**.

- Energy stored in fuels, food or batteries is called **chemical energy**.

- Energy stored in an object because it is lifted up is called **gravitational energy**.

- **Fuels**, food and **batteries** are all stores of chemical energy.

I3 Energy in food

Measuring energy

Energy is measured in **joules**, symbol **J**. One joule is the energy needed to lift an apple by one metre. A joule is very small so we use **kilojoules** instead. The abbreviation for kilojoule is **kJ**.

1 kilojoule = 1000 joules

Energy from food

Without food we would starve and die. Food is the fuel for our bodies. Food is a store of chemical energy.

We get fat if we eat too much food and take too little exercise. We get thin if we eat too little food. The amount of energy we need from food depends on what we do.

The bar chart below shows how much energy you need to do different tasks for one hour.

> **Did you know?**
>
> The joule was named after James Prescott Joule. James never went to university. He started investigating energy in a small laboratory in his father's brewery. By the age of 31 he was a famous scientist.

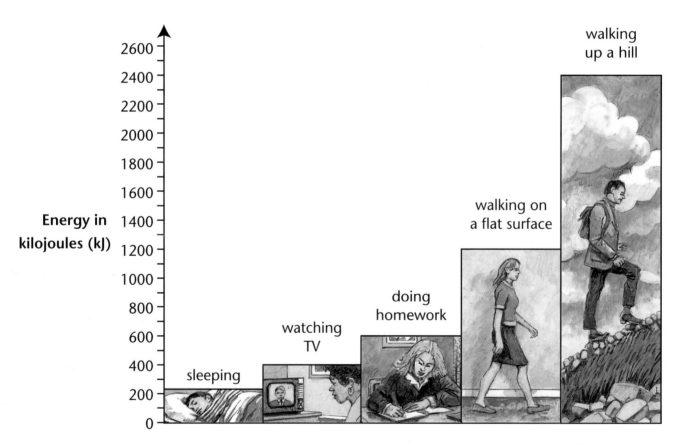

a How many kilojoules do you use to sleep for one hour?

These pictures show how much energy is in some different foods.

sausage 500 kJ   bread and butter 600 kJ   chips 1000 kJ   fried egg 150 kJ   tea with milk 65 kJ

b How much homework can you do using the energy in one slice of bread and butter?

Telling how much energy is in food

Processed foods have a label that tells us about the food. One of the things it tells us is the amount of energy in the food.

c How much energy is there in half a chicken and leek pie?

Nutritional information from a chicken and leek pie

Average values	per 100 g	per half pie
Energy	1050 kJ	1975 kJ

d Frank doesn't want to get fat. Look at the pictures of the foods above. Make up a breakfast for Frank that contains 1265 kJ.

Where does the energy come from?

Our food comes from plants and animals. Look at the diagram. The plants get their energy from the Sun. Animals get their energy from plants.

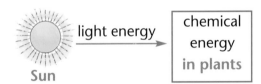

Questions

1 Copy and complete these sentences using the words below.

chemical   Sun   joules   kilojoule

We measure energy in _____. One thousand joules equals one _____.

Energy is stored in food as _____ energy.

The energy in our food comes from the _____.

2 Use the information on these two pages to answer these questions.

a Which needs more energy, walking on the flat or walking up a slope?

b How much energy would you take in if you ate a sausage and a portion of chips?

3 Fred eats an apple. Where did the energy stored in the apple come from?

For your notes:

- Energy is measured in **joules, J** and **kilojoules, kJ**. 1 kilojoule = 1000 joules.

- Food contains energy that originally came from the Sun.

- We take in chemical energy in our food to give us energy for life.

93

Investigating fuels

Shaibal's class investigated the energy given out by two different fuels, lighter fuel and firelighters. The class tested each fuel in turn.

They burned the fuels and heated water with them. The fuel that heated up the water more, gave out more energy.

a The class decided to use the same mass of fuel each time. Why do you think they did this?

b They decided to heat the same volume of water with each fuel. Why do you think they did this?

The pictures below show Shaibal and Pippa's experiment with lighter fuel.

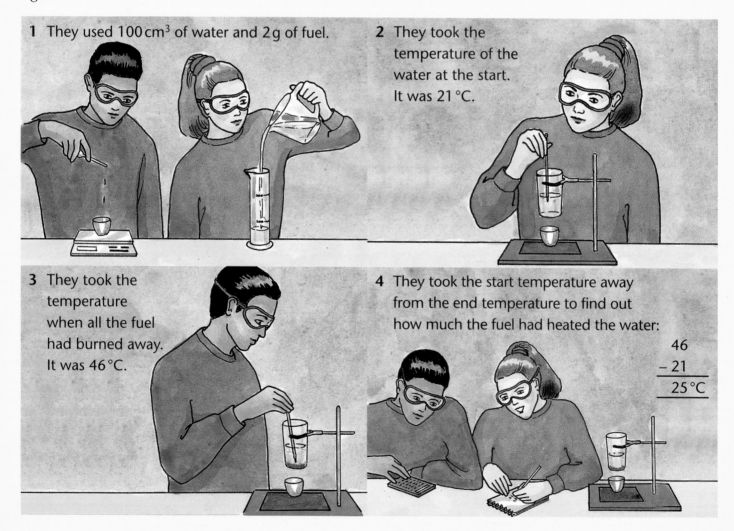

1 They used 100 cm³ of water and 2 g of fuel.

2 They took the temperature of the water at the start. It was 21 °C.

3 They took the temperature when all the fuel had burned away. It was 46 °C.

4 They took the start temperature away from the end temperature to find out how much the fuel had heated the water:

```
   46
 − 21
 ─────
 25 °C
```

They then did another experiment using firelighters instead of lighter fuel. They wanted to compare the firelighters with the lighter fuel.

Shaibal and Pippa burned 5 g of firelighter and heated 100 cm³ of water. The temperature of the water was 21 °C at the start and 82 °C at the end.

c How much did the fuel heat the water?

d Did they use the same mass of fuel as before?

e Did they use the same volume of water as before?

f Do you think this was a fair test? Explain your answer.

g Would you have done the experiment in the same way as Shaibal and Pippa? Explain your answer.

Variables

There are three things that could be different at the start of this investigation. They are called **variables**. They are:

● the type of fuel ● the mass of fuel ● the volume of water.

The class are investigating types of fuel so this is the only thing they change. It is called the **input variable**.

The change in water temperature depends on the type of fuel being used. This is called the **outcome variable**. This is the variable that they measure.

Shaibal and Pippa then did another investigation. They used lighter fuel for all their experiments. Their results are shown in the table below.

Mass of fuel in g	Amount of water in cm³	Temperature of water at start in °C	Temperature of water at end in °C	Temerature rise in °C
1.0	100	21	33	12
1.5	100	21	38	17
2.0	100	21	46	25

h What variable did they change? (This is the input variable.)

i What did they measure to get their results? (This is the outcome variable.)

j What variables did they keep the same to make it a fair test?

What are fossil fuels?

We find **coal**, **oil** and **natural gas** inside rocks. They are also called **fossil fuels**. A **fossil** is the dead remains of an animal or plant found in rock. A fuel is a material that gives out energy when it is burned.

a What are coal, oil and natural gas called?

How did fossil fuels form?

1 Millions of years ago, plants trapped the energy in sunlight, made food and grew.

2 When the plants died they were buried. The plants did not rot. This was because they were away from the air. Oxygen in the air helps rotting.

3 More layers of mud and sediment pressed down on the plants.

4 This turned the plants into coal and the mud and sediment into rocks. This took millions of years.

b Why did dead plants turn into coal rather than rotting away?

Oil and natural gas were made in a very similar way. Oil and natural gas were made from tiny sea animals rather than plants.

c The tiny sea animals ate plants. Look at the diagram below. How did these plants make their food?

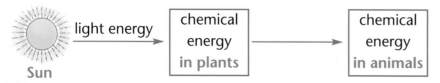

Using fossil fuels

Most of our fossil fuels are burned in power stations to make electricity. Power stations turn thermal (heat) energy into electrical energy.

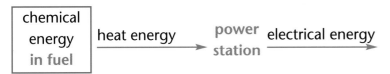

We also use fossil fuels to heat our homes and move our cars, lorries and planes. We can make useful materials like plastics from oil.

d List three uses of fossil fuels.

Non-renewable energy resources

Fossil fuels are going to run out. They are **non-renewable energy resources**. The fossil fuels took many millions of years to form. They are not being replaced as we use them.

e Look at the bar chart. How many years will it be before we use up all the:
(i) coal?
(ii) natural gas?
(iii) oil?

Estimated date when fuel will run out

2240 — coal
2200 —
2160 —
2120 —
2080 — natural gas
2040 — oil
2000 —

Questions

1 Copy and complete these sentences using the words below.

> Sun   non-renewable   natural gas
> animals   coal   oil   plants

Three fossil fuels are _____, _____ and _____ _____.

Fossil fuels were made from _____ and _____ that lived many millions of years ago. The energy in them came from the _____.

Fossil fuels will run out one day. This is because they are _____.

2 Which of statements **A** to **D** are true and which are false?

A The energy stored in oil came from the Sun.

B Fossil fuels could be renewed within 10 000 years.

C Oxygen is needed for fossil fuels to form.

D Coal was made mainly from dead plants.

3 Why will there probably be no natural gas or oil in 60 years' time?

For your notes:

Fossil fuels:

● include **coal, oil** and **natural gas**

● were made from dead animals and plants that lived many millions of years ago

● contain energy that came from the Sun

● are **non-renewable energy resources** that are running out.

97

I6 Using fuels wisely

Learn about:
- Conserving fossil fuels
- Renewable fuels

Making it last

Fossil fuels are non-renewable. They will run out. They will last longer if we use them more carefully. This is called **conserving** fossil fuels.

Everyone needs to stop wasting energy.

- We need to share cars rather than driving everywhere alone.

- We need to drive cars with smaller engines.

- We need to insulate our homes properly, so less thermal (heat) energy escapes. Then we burn less fuels heating our homes.

- We need to turn off lights to save energy. Then we use less electricity and that means the power station burns less fuel.

a Give three ways of saving energy, so that fossil fuels are conserved.

Biomass is renewable

Fossil fuels are running out. We need to find **alternative energy resources**. We can burn plant and animal material as fuel. This material is called **biomass**. We can burn biomass to heat our homes, as fuels for our cars and to make electricity.

SAVE ENERGY!

Switch off — Turn off lights when you leave a room. Turn off TVs, computers, and radios when you're not using them – many constantly use small amounts of power.

Light efficiency — Use energy efficient light bulbs wherever possible. They use less electricity and last up to six times longer.

Go green — Buy your home's electricity from a renewable energy source. See www.greenprices.com for more information.

Recycle — 80-90% of household rubbish ends up in landfill sites. Recycling saves landfill space, energy, raw materials and cuts air pollution.

- Use mains electricity wherever possible – or rechargeable batteries (they can be recharged up to 1000 times).
- Recycling glass and plastics conserves energy and oil reserves. 70% of plastics could be recycled – but only 1% currently is.
- Recycling aluminium cans instead of using raw materials can cut the energy needed by 90%.

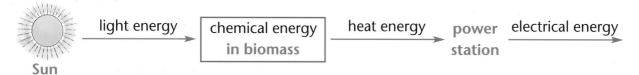

b Where does the energy stored in the biomass come from?

Wood has been used by humans as a fuel for many thousands of years. We could plant trees and harvest them. It would take 10 to 15 years to grow a tree. However, wood is still a **renewable energy resource** because we can replace it.

c Compare wood with coal. Why is wood renewable and coal non-renewable?

98

The Inuit people (Eskimos) in the photo below (left) live in the far north. They burn animal fat for fuel.

The Bedouin people live in the desert. They use camel dung as fuel.

d **Explain why the Inuits and the Bedouin don't use wood as a fuel.**

Wood is not the only type of biomass. Rotting material makes **methane**. This is the same substance that is in natural gas. Any plant or animal material could be rotted to make methane. The tank in the photo on the right contains farm waste that is being used to make methane.

e **Why is it a good idea to use rubbish and waste to make methane?**

Questions

1 Copy and complete these sentences using the words below.

> **wood   non-renewable   renewable**
> **methane   conserve   living things**

Fossil fuels are _____, so they are running out. We need to _____ fossil fuels to make them last longer.

Biomass is material from _____ _____. It can be burned instead of fossil fuels. Biomass fuels are _____. We can get _____ from trees and _____ from rotting rubbish.

2 Explain how the following will help us conserve energy:

  a sharing cars   **b** driving cars with smaller engines

  c insulating our homes **d** switching off lights.

For your notes:

- We need to **conserve** fossil fuels so that they last longer.

- **Biomass** is plant and animal material. It is a store of chemical energy.

- Wood and **methane** are examples of biomass that can be used as fuel.

- Many forms of biomass are **renewable energy resources** because the plants and animals grow quickly.

17 More energy resources

Solar energy

There is a lot of thermal energy in sunlight – you can burn paper using a magnifying glass! Look at this photo. **Solar furnaces** are huge, curved mirrors that concentrate sunlight like a magnifying glass. The heat (thermal) energy can then be used to make electricity.

Sun → heat energy → power station → electrical energy →

The Sun shines every day, although clouds sometimes cover it. **Solar energy** is a renewable energy resource.

ⓐ **We do not build solar furnaces in Britain. Why not?**

Wind energy

We can use the wind to make electricity. Look at the **wind turbine** in the photo on the left. The box at the top converts the movement of the sails into electrical energy.

The Sun makes the wind. Thermal energy from the Sun heats the air. The air moves and there is wind. The Sun heats the air every day, so **wind energy** is a renewable energy resource.

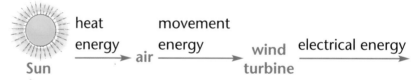

Sun → heat energy → air → movement energy → wind turbine → electrical energy →

ⓑ **Where would you build a wind turbine?**

Wave energy

Wave energy can also be used to make electricity by turning a **wave turbine**. Wind passing over the water makes the waves. The Sun heating the air makes the winds. This means that the energy in the waves comes from the Sun, so waves are a renewable energy resource.

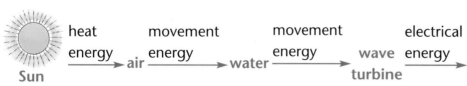

Sun → heat energy → air → movement energy → water → movement energy → wave turbine → electrical energy →

ⓒ **What makes the waves?**

Falling water

We use falling water to make electricity. The water is high up in a reservoir, so it is a store of gravitational energy.

When the water falls it has movement (kinetic) energy. A **water turbine** takes in the movement energy and gives out electrical energy. Look at the photo. The water falls down through the dam, making electricity.

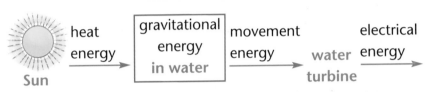

The Sun evaporates water from the seas. When it rains in the mountains, some water ends up in reservoirs high up. So it is energy from the Sun that lifts up the water.

The Sun shines every day, so falling water is a renewable energy resource.

d **Energy is needed for water to get up behind the dam. Where does this energy come from?**

No burning

There is another advantage of using solar energy, wind energy, wave energy or falling water. They do not involve burning. Burning fuels makes waste gases, which pollute the air.

Did you know?

This car travelled all the way across Australia without burning any fuel. The car converts solar energy to movement energy.

Questions

1 Copy and complete this table by ticking the correct boxes.

Energy resource	Is it renewable?	Is the energy from the Sun?	Does it make waste gases by burning?
fossil fuels			
biomass fuels			
solar energy			
wind energy			
wave energy			
falling water			

2 What advantages do renewable energy resources have over non-renewable energy resources?

3 Which energy resources do not add to air pollution?

For your notes:

- **Alternative energy resources** include **solar energy, wind energy, wave energy** and falling water.

- These are all **renewable energy resources**.

- They all get their energy from the Sun.

- They do not involve burning, so they don't make waste gases.

J1 Electrical energy

Electricity everywhere

You know that electricity can light our homes and streets and warm our houses. It can also make our washing machines, TVs, microwave cookers and computers work. But how does it do all this?

Look at the diagram on the right. **Cells** store **chemical energy** and transfer **electrical energy** to the circuit. Electricity carries energy to make things work.

a **What transfers energy to the circuit?**

The diagram below shows the energy transfer that happens in the circuit above.

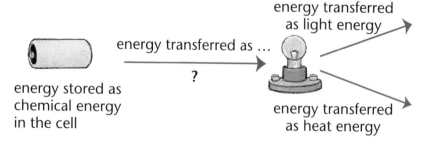

energy stored as chemical energy in the cell

energy transferred as ... ?

energy transferred as light energy

energy transferred as heat energy

b **What do we call the energy that is transferred by electricity?**

Voltage tells us how much energy the electricity is carrying. The bigger the number of **volts**, the more energy the cell can transfer. The short way to write volts is **V**.

Do you remember?

A circuit needs an **energy** source, such as a cell. You need a **complete circuit** to make a lamp light.

The energy must be transferred along the wires to the lamp.

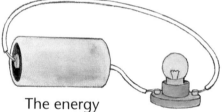

The energy comes from the cell.

The lamp transfers light energy and heat energy.

Do you remember?

We can use simple symbols to show cells, lamps and switches and how a switch works by breaking the circuit.

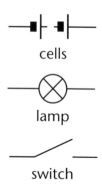

cells

lamp

switch

The number of volts is called the voltage of the cell.

C **Does a cell with 12 V store more energy than a cell with 1.5 V?**

Used cells have lost most of their stored energy so they won't make things work.

Cells or batteries?

You can place two or more cells together to put more energy into the circuit. In science, this is called a 'battery'. If you are connecting cells in a circuit make sure you connect the positive end of each cell to the negative end of the next one.

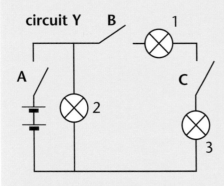

Questions

1 Copy and complete the sentences using the words below:

> **battery   electrical   chemical   cell   negative**

_____ energy from the cell is transferred into the circuit as _____ energy.

More than one _____ connected together makes a _____.
In a circuit, you must connect the positive end of each cell to the _____ end of the next one.

2 Look at circuit Y. All the switches are shown open.

circuit Y

Switches closed	Lamps lit
A, B, C	1, 2, 3
none	
A, C	
A, B	
A	

a How many lamps are there in this circuit?

b How many switches are there in this circuit?

c Copy and complete the table.

For your notes:

- We get **energy** from electricity to make things work.

- You need a **complete circuit** for energy to be transferred.

- **Cells** store energy. More than one cell connected together is called a **battery**.

- Cells with a high **voltage** can transfer more energy.

103

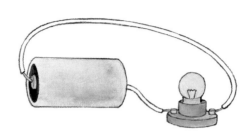

Current

In the diagram on the right, wires join the cell to the lamp. Electricity flows in the wires. We call this flow a **current**. The current flows around the whole circuit.

The electrical energy in this circuit is transferred as heat and light energy. It is transferred out of the circuit into the air and is replaced by more energy from the cell. But the current which carries the energy is never used up.

It helps to think about it like a central heating system:

● The water in the system is given heat energy by the boiler.

● The flow of water carries the heat energy to the radiators in the house.

● The heat energy is transferred to the radiators and into the air.

Just as the current in an electrical circuit is not used up, the water is the central heating system is also not used up. There is always the same amount of water in the system.

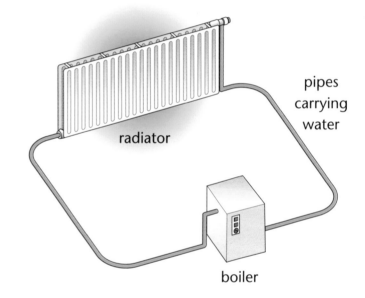

pipes carrying water

radiator

boiler

a **What part of the electrical circuit is like the boiler?**

b **What part of the circuit is like the pipes?**

c **What part of the circuit is like the radiator?**

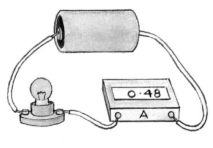

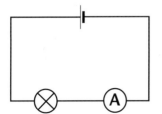

Measuring current

You can show that current is never used up by measuring it in different places in the circuit.

You can measure current by putting an **ammeter** into the circuit. The circuit symbol for an ammeter is (A).

An ammeter measures current in **amps**. The short way of writing amps is **A**.

d **Where do you put the ammeter to measure the current?**

e **What is current measured in?**

Moving the ammeter

You can move the ammeter to different parts of the circuit.
The teacher asked Ben, Laura and Dan what would happen.

Does it matter where you put the ammeter?

It will have to be in the circuit. Let's put it to the left of the lamp.

Ben

It doesn't matter where you put it. The current is the same on both sides of the lamp.

Laura

No, energy leaves at the lamp. I think the current will be different on either side. Let's use two ammeters to check.

Dan

f Who do you agree with?

They did the experiment to check their prediction. It is shown here.

g Who was right?

The current is the same on both sides of the lamp. This shows that the current is not used up as it travels around the circuit.

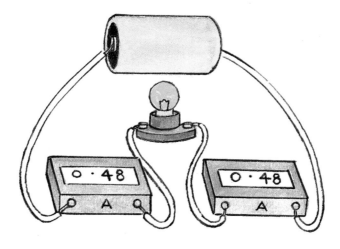

Questions

1 Copy and complete these sentences.

   There is a current in the _____. We measure current using an _____ . Current is measured in _____. The current is the _____ before and after a lamp.

2 Why doesn't it matter where you measure the current in a circuit like the one shown on this page?

3 How does energy get from a cell to a lamp in a circuit?

For your notes:

- **Current** flows around a circuit. It is measured in **amps, A**, using an **ammeter**.

- The current is the same on both sides of a lamp.

- Current is not used up as it flows around a circuit.

Flowing through

If you think about water flowing through pipes, it flows more easily through wide pipes than through narrow ones. Narrow pipes slow the flow down.

Thin and thick wires in a circuit are like narrow and wide pipes. A thin wire slows the current down because it is harder for the electricity to get through. The thin wire resists the electricity more so we say it has a high **resistance**.

An ordinary light bulb has a very thin wire in it which has high resistance to the current.

Components like light bulbs slow the current down.

Series circuits

In **series circuits** the lamps are arranged side by side in the same loop as shown here. The more lamps there are, the more the current is slowed down through the whole circuit.

The lamps shine less brightly than if there was only one lamp.

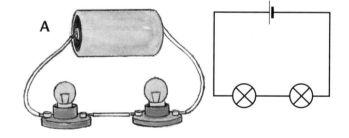

A

Parallel circuits

You can connect several lamps to the same size cell but keep them as bright as just one would be. You can put them in different loops. Look at the diagram on the right. This is called a **parallel circuit**.

The diagram shows the current in a parallel circuit. The current branches off and goes through the two bulbs at the same time, not one after another. So, the current is not slowed down twice.

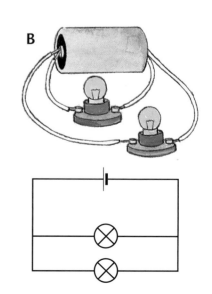

B

a Which circuit has the lamps side by side?

b Which circuit has the lamps in different loops?

c Which lamps are brighter?

Current in series and parallel circuits

Kimberly and Jason built these circuits. They measured the current at different points in each circuit.

In the series circuit, the current was the same at all points in the circuit. In the parallel circuit, the current was shared between the loops of the circuit.

d Look at the series circuit below. What would the current be at X and at Y?

e Look at the parallel circuit below. What would the current be at P, at Q and at R?

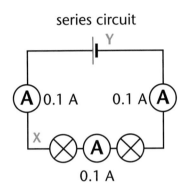

series circuit

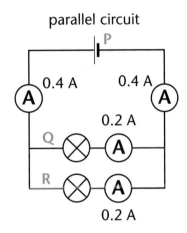

parallel circuit

Questions

1 Find the word that means the electric current finds it hard to flow through.

2 Read these statements.

- The lamps are brighter.
- The lamps are dimmer.
- The current is shared between the loops of the circuit.
- The current is the same at all points in the circuit.

Make a table and put each statement in the correct column.

Series circuit	Parallel circuit

3 What would happen to the current in a series circuit if a lamp with a higher resistance were used?

For your notes:

- **Resistance** makes it hard for the current to flow.

- You can connect lamps in **series** and in **parallel**. **Parallel circuits** have more than one loop.

- Two lamps in parallel are brighter than the same two lamps in series, using the same cell.

- In a **series circuit** the current is the same at all points.

- In a parallel circuit the current is shared between the loops.

J4 Models of electricity

Using models

Scientists use **models** to help them think. Each part of a model stands for something in real life.

A good model fits with the facts. So far, you know these facts about electricity.

> **Electricity carries energy to make things work.**

> **You need a complete circuit to make a lamp light.**

> **The current is the same on both sides of the lamp.**

We have used models of a central heating system and of water flowing through taps to talk about current in a circuit. We can use other models for electrical circuits.

The coal truck model

This model shows a mine and a power station. There is a single-track railway between the mine and the power station. Coal trucks run along the railway. The coal trucks can move quickly or slowly along the track.

At the mine, the coal trucks are filled with coal. The coal trucks run along the tracks and deliver the coal to the power station. The empty coal trucks then return to the mine.

Read the description carefully again and study the diagram. Then answer the questions.

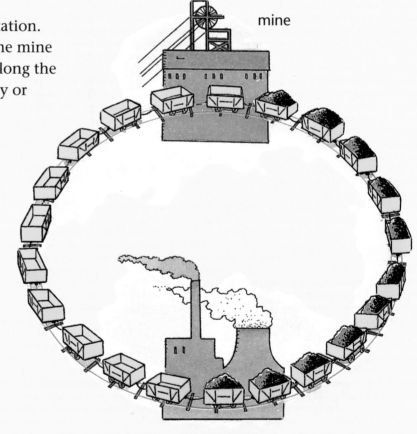

mine

power station

ⓐ In the coal truck model, what stands for:

    (i) the circuit?   (ii) the cell?
    (iii) the lamp?   (iv) the energy?

ⓑ The moving trucks represent the current. The trucks speed up. Has the current increased or decreased?

Models are very useful to explain how things work but the problem with a model is that it is never perfect.

In the coal truck model, if the coal trucks are derailed, coal will spill out onto the tracks.

c **What happens if there is a break in an electrical circuit? Does electricity spill out?**

The class and matches model

Mrs Fuller is using another model to explain electricity to her class.

d **Think about this model. What stands for:**
(i)   the energy?
(ii) the circuit?
(iii) the current?

e **Draw a diagram of this model. Use the same colours as in the coal truck model:**

● **energy is green**

● **the circuit is pink**

● **the current is yellow**

● **where energy goes into the circuit is blue.**

Mrs Fuller gives each pupil a match as they pass her.

The pupils carry their matches round the white circle.

The pupils continue and collect another match.

Mrs Huxley strikes each match as the pupil passes it to her.

Questions

1 Jackie and Lester drew diagrams to show the class and matches model of electricity.

  a Does Jackie's diagram show what happens in a circuit? Explain your answer.

  b Does Lester's diagram show what happens in a circuit? Explain your answer.

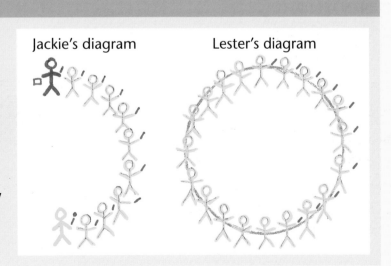

Jackie's diagram   Lester's diagram

J5 Electrical hazards

Finding a fault

A **fault** is something that stops an electrical circuit from working. You can find a fault in a circuit by replacing each part of the circuit until it works again.

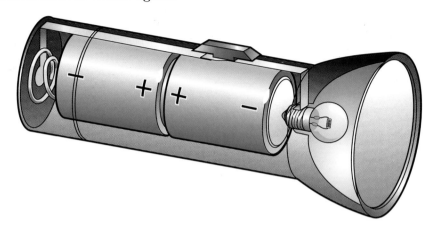

a **Look at the diagram above. What fault is stopping the torch from working?**

You can also find a fault by testing each part of the circuit to see if it is working properly. The best way to test a component is to put it in a circuit that you know is working.

Circuits in the home

The lamps in the rooms of a house are in a parallel circuit, if you switch off the lamp in one room, the others stay on.

Look at the diagram on the right; if you switch off the lamp in the kitchen, the lounge lamp stays on. If the lamps were in series, switching off one would switch off all of the others.

In homes, this type of circuit is called a **ring main**. One advantage of a ring main is that when one bulb does not work, all the others keep on working.

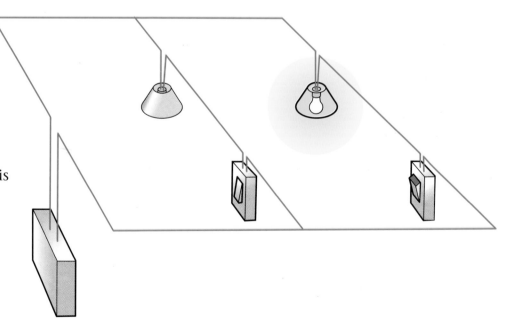

Why can electricity be dangerous?

Electricity from the mains supply delivers much more energy than the energy we get from the cells we use in radios, clocks, and cd-players. This is why we should be very careful when we use mains electricity at home.

A **fuse** is a safety device. All the plugs on the appliances we use have fuses.

If the current gets too high or there is a fault in the circuit, the fuse stops current flowing. This can help stop you getting an electric shock. It can also stop the wires getting hot and causing a fire.

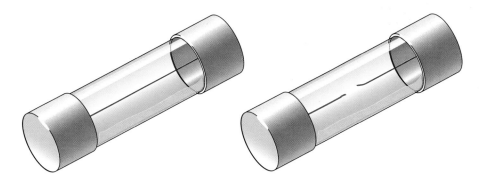

Fuses work because fuse wires have a high resistance. As the current increases, it has difficulty flowing through the fuse wire so energy is transferred as light and heat.

If the current gets too large, the heat causes the fuse wire to melt and break the circuit.

Do you remember?

If you get an electric shock from mains electricity (240 V), it can stop your heart beating.

Questions

1 Give two reasons why a parallel circuit is used to connect the lamps in a house.

2 You have two lamps and three switches. You want to make a parallel circuit that will:

   ● switch off both lamps together

and also

   ● switch off each lamp separately.

Draw the circuit that will do this.

3 Explain how a fuse can save lives.

For your notes:

● We can find a **fault** by testing each part of the circuit.

● The kind of circuit we use in our homes is called a **ring main**.

● Electricity can be dangerous. It can stop your heart beating.

● **Fuses** work by breaking the circuit if the current is too high.

K1 Forces and gravity

Learn about:
● Gravitational attraction
● Weight
● Mass

Forces everywhere

Everything you do uses forces. You cannot see forces, but you can often see the effects of a force.

A force can change the shape of an object. A force can make an object move faster or slower. A force can make an object change direction.

'Push' and 'pull' are two types of forces. The magnetic attraction between a magnet and iron is another type of force.

Do you remember?

You have used a forcemeter (newtonmeter) to measure forces. The '**newton**' is the unit for measuring force.

(a) **What units do we use to measure forces?**

What is gravitational attraction?

If you drop something it is pulled down to Earth by a force called **gravitational attraction (gravity)**. It is pulled towards the centre of the Earth.

The picture shows the Earth is shaped like a ball. Britain and Australia are almost on opposite sides. Gravitational attraction pulls Sharon and Shirley towards the Earth. The force acts down towards the centre of the Earth.

(b) **Which way is 'downwards' in Australia?**

(c) **Why does Shirley not fall off Australia?**

What is weight?

Samson has to pull up against a force to pick up the dumb-bell. This force is the **weight** of the dumb-bell, which pulls it down. Weight is caused by gravity on an object.

We measure weight in units called newtons or **N**. The weight of the dumb-bell is 300 newtons or 300 N.

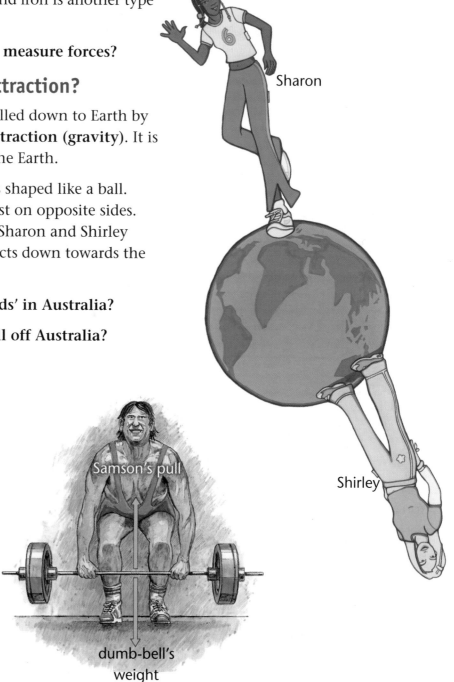

Sharon

Shirley

Samson's pull

dumb-bell's weight

Mass and weight

Sharon is made of a certain amount of stuff or **matter**. Sharon's **mass** is a measure of how much matter she is made of. Mass is measured in **kilograms** or **kg**. Sharon has a mass of 66 kg.

Sharon's weight is different. Her weight is caused by gravitational attraction pulling on her mass. On Earth, gravitational attraction has a force of 10 N on each kilogram. To find the weight of something you multiply its mass by 10. So, Sharon has a weight of $66 \times 10 = 660$ N.

Make sure you use the words mass and weight correctly in science.

d **If Shirley's mass is 45 kg, what is her weight on Earth?**

The bigger the mass of an object, the bigger its weight. We use weight to mean how 'heavy' something is. So heavy objects are pulled down with a bigger force than light objects.

e **Which of the weights will have the biggest pull?**
10 N   20 N   100 N

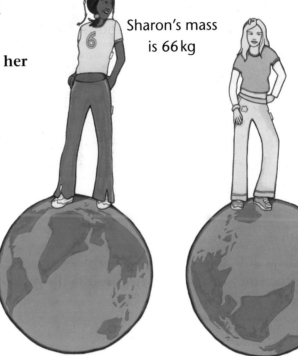

Sharon's mass is 66 kg

Shirley's mass is 45 kg

Questions

1 Copy and complete these sentences by choosing from the words below.

> **kilograms   gravitational attraction   matter**
> **newtons   same   weight**

Weight is the force of _____ on an object. We measure weight in _____.

Mass is a measure of how much _____ an object is made of. Mass is measured in _____.

2 Calculate the weight on Earth of these people:

  a Susan, mass 70 kg   **b** Philippa, mass 55 kg

  c Marco, mass 88 kg.

3 What is gravitational attraction?

For your notes:

- **Gravitational attraction** is the force that pulls an object and the Earth towards each other.

- **Weight** is the force of gravitational attraction on an object. We measure weight in **newtons, N**.

- **Mass** is a measure of how much **matter** an object is made of. We measure mass in **kilograms, kg**.

Friction

What is friction?

Friction is a force that acts when things rub together. Friction can slow things down.

The ice skates have very little friction on the smooth ice, so you can skate fast.

The runner's shoes have good grip and make lots of friction with the rough ground to help her slow down and stop.

ⓐ What kind of surfaces make the most friction?

ⓑ What kind of surfaces make the least friction?

Friction can be useful

Friction can be a very useful force. Bikes and cars have brakes that use friction to slow them down or stop them. The surfaces of the brakes rub against the wheels so the wheels don't turn so fast.

ⓒ Think of another example showing how friction is useful in everyday life.

Reducing friction

Sometimes friction is not useful and we want to reduce it. When two surfaces rub together, they will become worn down because of friction.

Machines have a lot of parts that rub together. To reduce friction, we use oil and grease. We call these **lubricants**. They make surfaces run smoothly against each other.

d What would you do to make the chain on your bike run smoothly?

Friction makes things warm

Where there is friction, heat energy is given out. You can feel this happen when you rub your hands together. They feel warm.

Elijah McCoy

Elijah McCoy made an important invention. Not many people know of his work. He was born in Ontario, Canada in 1844. Elijah's parents escaped from slavery in the USA. He studied engineering in Scotland and then returned to work in America.

Elijah worked on the railroads oiling the engines of trains. In his spare time, he made a cup that sent oil automatically to the engine, so that the train did not have to stop to be oiled. It was called the 'Real McCoy Lubricating Cup'.

Elijah McCoy and his invention.

Questions

1 Copy and complete these sentences by choosing from the words below.

   freeze grease heat light lubricants rub tyres

   Friction is made when two surfaces _____ together.

   Where there is friction, _____ energy is given out.

   Friction can be reduced by using _____ such as oil and _____.

2 Write a story about a world without friction.

3 Describe Elijah McCoy's invention.

For your notes:

● **Friction** is a force that acts when things rub together.

● We can reduce or increase friction to make it useful to us.

● We use **lubricants** to reduce friction between moving parts.

K3 Balanced forces

Staying put

Look at the picture of Zena and Sam having a tug-of-war. They are not moving. They are pulling with the same sized force, but in opposite directions.

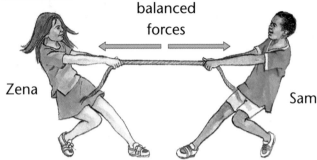

balanced forces

Zena   Sam

The forces are shown with **force arrows**. A force arrow points in the direction of the force. The length of the arrow shows the size of the force.

If two forces are the same size and pull in opposite directions, the forces are **balanced**.

a **Why might an object stay where it is, even when there are forces acting on it?**

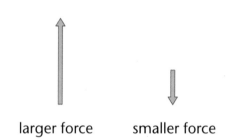

larger force   smaller force

More balanced forces

The picture shows a mass hanging from a spring on a forcemeter. The amount the spring stretches is called the **extension**. The mass is not moving. The forces on it are balanced.

Weight is the force pulling down on the spring with the same force as the spring pulls up on the mass.

Mr Blue the decorator is standing on a plank. The plank bends down because of Mr Blue's weight. The plank pushes up. This force from the plank is called a **reaction force**.

The force pushing down on the plank is the same as the force pushing up on Mr Blue. The forces are balanced. If they were not, the plank would break.

b **What are the forces on your chair when you sit still on it? Draw a diagram with arrows.**

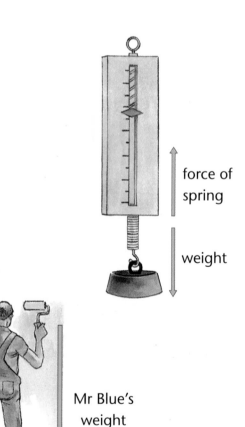

force of spring

weight

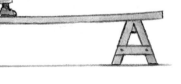

reaction force of plank

Mr Blue's weight

116

Why do things float?

Look at the two photos. Polystyrene **floats**. If you push it down under the water, you can feel it pushing up against your hand. This force is called **upthrust**. The object pushes down on the water. This force is called weight.

If the object floats, the upthrust is equal to the weight. The forces are balanced.

There are also balanced forces on a hot-air balloon floating in air.

c What is the name of the force that pushes up on the balloon?

d What is the name of the force that pushes down on the air?

Questions

1 Copy and complete these sentences by choosing from the words below.

> balanced   gravitational attraction   reaction
> weight   upthrust   unbalanced

When two forces are equal and in opposite directions, they are called _____ forces.

The force from the plank when a decorator stands on it is called a _____ force.

If a man pulls a dog with a force of 10 N, and the dog pulls the man with a force of 10 N the forces are _____.

The forces of _____ and _____ are balanced when a hot-air balloon floats.

2 Gianni pulled his dog with a force of 40 N and the dog pulled back against him with a force of 40 N. Draw a diagram of this and say why the forces are balanced.

3 Explain why a boat floats on water.

For your notes:

- **Force arrows** on a diagram show the direction of a force.

- If two forces are the same size and pull in opposite directions, they are called **balanced forces**.

- The **reaction force** stops something falling through a solid object. The reaction force balances the weight.

- When an object **floats**, the forces of weight and **upthrust** are equal.

117

Unbalanced forces

The box in the picture has two forces acting on it. One is the pull of the rope. The other is the weight of the box.

The force arrows are different lengths. This tells us that the forces are different sizes. They are called **unbalanced forces**.

ⓐ Which force is bigger?

ⓑ In which direction do you think the box will move?

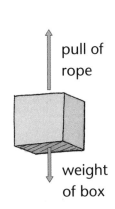

pull of rope

weight of box

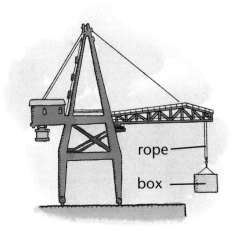

rope

box

Getting going

If Dipal does not push his go-kart, it will not start moving. A force is needed to start something moving.

Dipal gave his go-kart a gentle push. It did not move at all. Friction stopped it moving.

Dipal gave his go-kart a bigger push. Dipal's push on the go-kart was bigger than friction, so the go-kart started to move.

Dipal's push

friction

When forces push against each other like this, and one force is bigger than the other, they are called unbalanced forces.

When there are unbalanced forces acting on an object, the object starts to move. It moves in the direction of the bigger force and it gets faster.

ⓒ What will happen to the boxes A and B?

A

B

Shaping up

Unbalanced forces can also change the shape of an object. It might become bent, twisted or even break.

d What do you think would happen if unbalanced forces act on a big foam mattress?

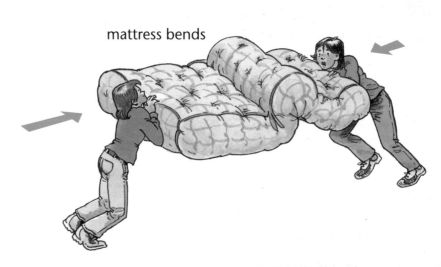
mattress bends

Unbalanced forces on moving objects

Unbalanced forces can act on something that is already moving. The car in the diagram is moving forwards.

Air makes friction with moving objects such as cars and planes. We call this **air resistance**. Because the force from the engine is bigger than the air resistance, the car moves faster.

When the bigger force is in the same direction as the moving object, the object speeds up.

When the bigger force is in the opposite direction to the moving object, the object slows down.

Do you remember?

Air resistance is a force that slows down objects moving through air. Try walking into the wind!

force of engine   air resistance

Questions

1 Copy and complete these sentences by choosing from the pairs of words below

   bigger/smaller   speeds up/slows down   move/stop

   When there are unbalanced forces acting on an object, the object starts to _____.

   It moves in the direction of the _____ force.

   When the bigger force is in the same direction as a moving object, the object _____ _____.

   When the bigger force is in the opposite direction to a moving object, the object _____ _____.

2 Draw arrows to show the following forces. Use 1 cm to show 1 N. So you would draw an arrow 2 cm long to show a force of 2 N.

   a 1 N   **b** 5 N   **c** 10 N   **d** 4 N.

For your notes:

- **Unbalanced forces** can act on an object that is not moving. The object starts to move in the direction of the bigger force.

- Unbalanced forces can also change the shape of an object. It may bend, twist or break.

- Unbalanced forces can act on a moving object. They can make the object speed up or slow down.

119

K5 Slow down!

Talking about speed

People use different sayings to describe how fast or slow things move.

We can tell how fast a thing moves by measuring its **speed**.

As fast as the speed of light.

As swift as a deer.

Slower than a snail.

Faster than a speeding bullet.

How do we measure speed?

To find the speed of an object, you need to know the distance the object travels and the time it takes to travel that distance.

We measure distance in metres or kilometres, and time in seconds or hours. In science, we measure speed in **metres per second** or **m/s**. In everyday life, we find it easier to measure speed in **kilometres per hour** or **km/h**.

If a car travels 60 km in 1 hour, its speed is 60 km/h.

Speed check

The picture on the right shows the speeds of some moving objects.

ⓐ Which is the fastest animal?

ⓑ At what speed does it travel?

ⓒ If a dog ran at a speed of 15 metres per second or 15 m/s, how far would it travel in 1 second?

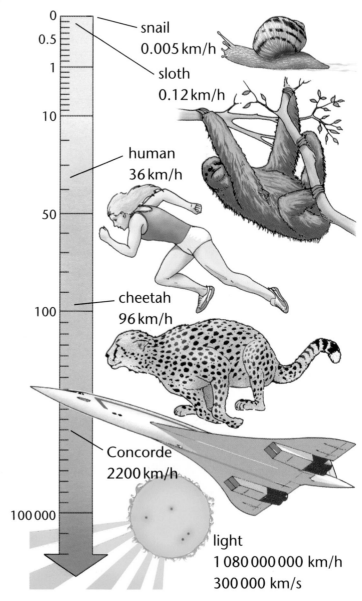

snail 0.005 km/h
sloth 0.12 km/h
human 36 km/h
cheetah 96 km/h
Concorde 2200 km/h
light 1 080 000 000 km/h
300 000 km/s

Stopping distances

When a driver decides to stop, it takes less than a second to step on the brakes. The car travels several metres in this time. This is called the **thinking distance**.

After the driver steps on the brake, the car slows down. But it continues to travel further until it stops. This is called the **braking distance**.

The overall **stopping distance** is the thinking distance + the braking distance.

The faster a car is travelling, the longer it takes to stop and the further it travels while it is stopping.

30 mph

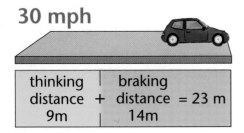

| thinking distance 9m | + | braking distance 14m | = 23 m |

60 mph

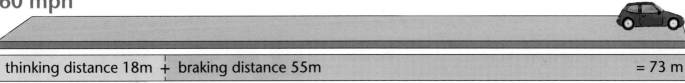

| thinking distance 18m + braking distance 55m | = 73 m |

The picture shows stopping distances for a car with good tyres in good weather. If the tyres grip the road there will be more friction to slow down the car. If the tyres are worn, or the road is wet, there is less friction to slow down the car.

(d) How do we work out the overall stopping distance for a car?

(e) What happens to the stopping distance as a car moves faster?

Questions

1 Copy and complete these sentences by choosing from the words below.

   distance   hour   minute   second   kilometres

   The speed of an object is usually measured in metres per _____ or _____ per hour.

   To find the speed of an object, you must find the _____ the object travels and the time taken for it to travel along.

2 How could you find out who is the fastest runner in your class?

3 Why does it take longer for a car to stop in wet weather?

For your notes:

- **Speed** is the distance an object travels in a certain time.

- The units used for speed are **metres per second, m/s,** or **kilometres per hour, km/h.**

- The faster a car is travelling, the longer it takes to stop.

K6 Archimedes' story

A new image

King Hiero reigned in ancient Greece. One day he decided that it was time for a new look. 'I know what I need – a brand new crown,' he thought. He sent for the court goldsmith and gave him a lump of gold to make the crown.

Crowning glory

Days later, the goldsmith delivered the crown. The next day, the King began to worry. There were rumours that the goldsmith was dishonest. The King suspected that the goldsmith had kept some of the valuable gold and put cheap silver in the crown.

a Which metal did the King think had been added to the crown?

Archimedes to the rescue

Hiero sent for his adviser, the brilliant young scientist Archimedes, to find out if the crown was real gold.

Back home, Archimedes started to think about this problem. Solving it would be hard because he could not damage the crown. He decided to have a bath. He did not notice that it was full to the brim! As he climbed in the water overflowed onto the floor.

b Why do you think the water overflowed?

That moment he had an idea. He leapt out of the bath and ran naked through the market shouting 'I have found it!'

Archimedes had realised that he could measure the amount of space an object takes up by measuring the amount of water it **displaces**. His body had pushed out, or displaced, the water that had overflowed onto the floor.

Archimedes burst into the throne room. 'Your Majesty! Have you got a lump of gold the same as the one you gave the goldsmith? I have a way of testing your crown.'

Obtaining evidence

Archimedes put the crown on the left-hand side of a balance and then put the lump of gold on the right-hand side. The scales balanced.

'So, my crown is all gold,' the King said, a little surprised.

c **Why do you think the King was surprised?**

'Wait,' said Archimedes as he filled a large washbasin with water. He carefully lowered the crown into the basin and collected the water that overflowed in a larger basin. He then did the same thing with the lump of pure gold.

'It's pushed out less water than the crown!' cried the King.

Evaluating the evidence

'The crown is not made of pure gold,' went on Archimedes. 'If it was pure gold, it would have pushed out the same amount of water. The goldsmith is a thief. He has used some silver in the crown and kept some gold for himself.'

Archimedes knew that silver is lighter than gold. We say that it is less **dense** than gold, so more of it would be needed to make the crown the same mass.

This would mean that the silver would take up more space and push out *more* water than the gold. This is what happened, so Archimedes knew the crown had silver in it.

d **What scientific knowledge did Archimedes use to help him explain his idea to the King?**

'Send for the goldsmith! He is not going to get away with this one,' shouted King Hiero.

Questions

1 Discuss the story of Archimedes with your partner. Copy and complete these sentences together by choosing from the words in bold.

Silver is **lighter/heavier** than gold. So to make the crown the same mass, **less/more** silver was needed. The crown took up **less/more** space and pushed out **less/more** water than the lump of gold because **less/more** silver was needed to replace the gold he took out.

2 Archimedes knew that there was only way to find out whether the crown was made of pure gold or not. A crown with silver in it would take up more space that the original block of gold.

How do you think he used his everyday experience to solve the problem of finding out how much space the crown took up?

L1 Shedding light

Learn about:
- Light
- Stars and planets
- Reflection

Seeing the light

Do you remember?

You already know that light travels from a source. A torch is a **light source**. When you shine it onto something, light is travelling from a source.

Look at the three photos below. All these objects give out light, which travels to us so that we can see them.

  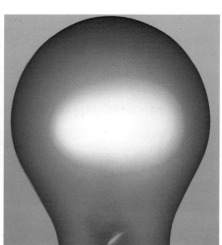

ⓐ **What does a source of light do?**

Light travels in a straight line, at a speed of 300 000 000 m/s. It is sometimes hard to understand just how fast light is. If you have ever been to America you will know that it takes a long time to fly there. Light is so fast that it gets there in less than a second!

The Sun as a star

Our Sun is a source of light for everyone on Earth. It is a star, and all stars are a source of light. It takes just 8.3 minutes for the light from the Sun to reach us. It is 150 million kilometres away.

ⓑ **Why is the Sun so important for everyone on Earth?**

The photo on the right shows some stars in the night sky. Like all stars, our Sun is a huge ball of very hot gas. The temperature at the surface of the Sun is about 6000 °C.

ⓒ **What is the Sun made of?**

Seeing stars

The Sun and stars are **luminous**, which means they give out light.

Most of the stars we can see are in our **galaxy**, the Milky Way. Our galaxy has 100 000 million stars. This photo of stars was taken by the Hubble telescope.

Seeing planets

Planets such as the Earth do not make their own light. They are **non-luminous**. We see planets only because light from the Sun bounces off them (is **reflected**) and reaches us. The diagram below shows this.

Do you remember?

We see light because light from a source enters our eyes.

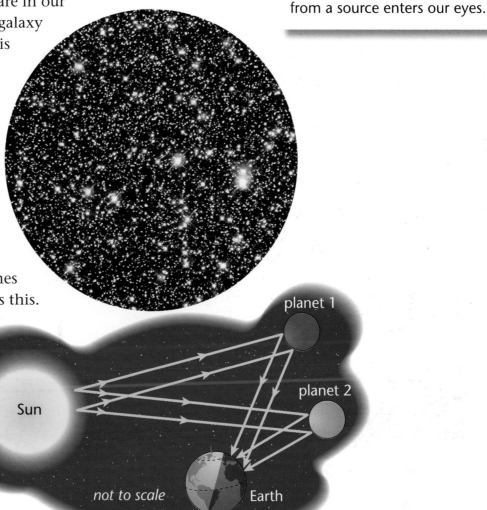

planet 1

planet 2

Sun

not to scale   Earth

Questions

1 Which of the following statements are true? Write out the ones that are true.

   a The temperature at the surface of the Sun is about 3000 °C.

   b The Sun is a planet.

   c Light takes 8.3 seconds to reach us from the Sun.

   d We see the planets because the Sun's light is reflected off them.

2 In which galaxy are the stars we can see?

3 Find the information on these two pages about how far the Sun is from Earth. The next nearest star after our Sun is Alpha Centauri, 40 000 billion kilometres away. Why do you think it takes longer for light from this star to reach us than light from the Sun?

For your notes:

● The Sun and stars are **light sources**.

● We see planets and moons because they **reflect** light from the Sun to us.

● Light travels in a straight line at 300 000 000 m/s.

● Most of the stars we can see belong to our **galaxy**, the Milky Way.

Learn about:
- Day and night
- Phases of the Moon
- Eclipses

Day and night

In the morning, the Sun appears to rise in the east. In the evening, it appears to sets in the west. But the Sun does not move at all. It only seems to move across the sky because the Earth is spinning.

When the Sun shines on the spinning Earth, only the side of the Earth facing the Sun gets light. The other half of the Earth is in shadow. As the Earth spins, the UK moves into the light. This gives us day and night.

a Look at the diagram. Is it day or night in the UK?

b What causes day and night?

c Does the Sun actually move? What is really happening?

Do you remember?

The Earth spins on its **axis** once every 24 hours. The axis is an imaginary line that runs from the North Pole to the South Pole.

Phases of the Moon

As the Moon orbits the Earth, its shape appears to change every day. These changes are called the **phases of the Moon**.

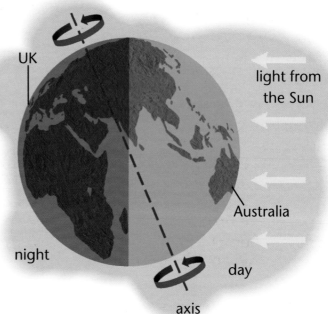

Do you remember?

The Moon goes round the Earth once approximately every 28 days. The path it takes is called an **orbit**.

day 1	day 4	day 7	day 11	day 14	day 18	day 21	day 25

new Moon   first quarter   full Moon   last quarter

d How long does the Moon take to orbit the Earth?

We see a full Moon when the Moon is on the opposite side of the Earth from the Sun (day 14). When the Moon is between the Sun and the Earth (day 1) no sunlight shines on the side we see from Earth. This is called a new Moon.

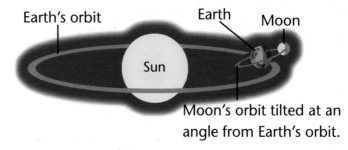

Earth's orbit   Earth   Moon

Sun

Moon's orbit tilted at an angle from Earth's orbit.

Lunar eclipse

Sometimes the Moon, Earth and Sun are lined up so the Earth blocks the light from the Sun. We see a shadow of the Earth on the Moon. This is called a **lunar eclipse**.

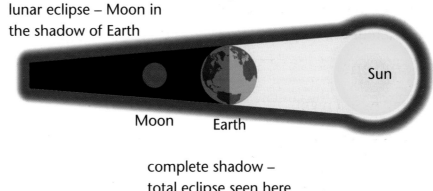

lunar eclipse – Moon in the shadow of Earth

Moon   Earth   Sun

Solar eclipse

Sometimes the Earth, Moon and Sun are lined up so that the Moon blocks the light from Sun. We see a shadow of the Moon on part of the Earth. This is a **solar eclipse**.

e What happens to the light from the Sun in a lunar and a solar eclipse?

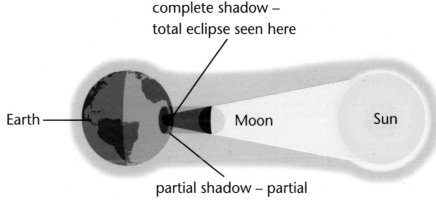

complete shadow – total eclipse seen here

Earth   Moon   Sun

partial shadow – partial eclipse seen here

Questions

1 Copy and complete these sentences to explain how day and night happen, using the words below:

24 hours   axis   day   night

The Earth spins on its _____ once every _____.

When the Sun shines on the Earth, on the side facing the Sun it is _____ time.

On the side not facing the Sun, it is _____ time.

2 What would happen if the Earth were to spin more slowly on its axis?

3 Draw a diagram to show the positions of the Sun, Earth and Moon during:

a a lunar eclipse   **b** a solar eclipse.

For your notes:

- The Earth spins on its **axis** approximately once every 24 hours, to give day and night.

- The Moon **orbits** the Earth once every 28 days, to give the **phases of the Moon**.

- Sometimes the Earth is between the Sun and the Moon, so we see a **lunar eclipse**.

- Sometimes the Moon is between the Sun and the Earth, so we see a **solar eclipse**.

L3 All in a year

Earth years

The Earth actually orbits the Sun every 365¹/₄ days. Every four years we have a **leap year**. A normal year has 365 days, while a leap year has 366 days. The extra day comes from adding together the extra four quarters of a day every four years.

The seasons

During the year, the climate in the UK changes. We have four **seasons**: spring, summer, autumn and winter.

a **Describe how the temperature in the UK changes with the seasons.**

The axis the Earth spins on is slightly tilted. The UK is on the top half of the Earth, called the **northern hemisphere**. The tilted axis gives us the seasons.

b **What gives us the seasons?**

When the northern hemisphere is tilted away from the Sun it is winter in the UK, and when it is tilted towards the Sun it is summer.

When the UK is tilted away from the Sun it gets less energy from the Sun, so it is colder. When the UK is tilted towards the Sun it gets more energy, so it is warmer.

Do you remember?

As well as spinning on its axis, the Earth orbits the Sun once every 365 days. This is called a **year**.

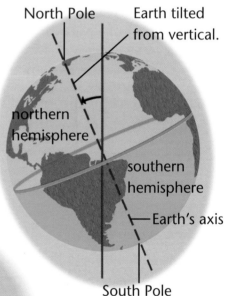

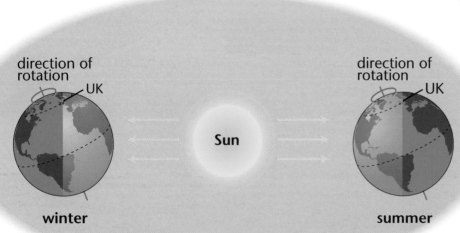

c **Which way is the Earth tilted to give us warm summer days in the UK?**

Daylight hours

In midsummer we have long days and short nights. The longest day is 21 June and the Sun is highest in the sky. It is above the horizon for the longest time.

In midwinter we have short days and long nights. The shortest day is 21 December and the Sun is lowest in the sky. It is above the horizon for the shortest time.

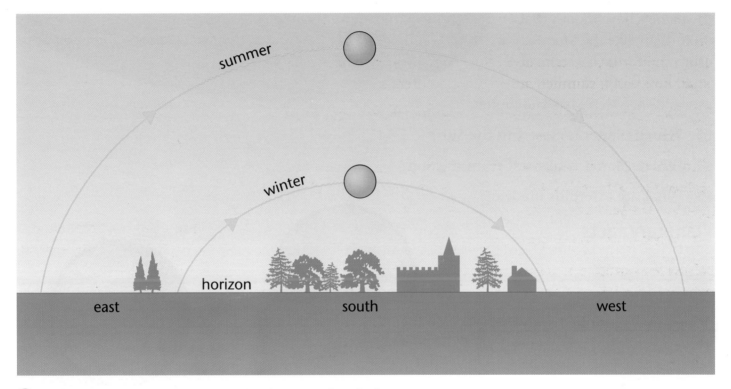

d In which season is the Sun highest in the sky?

Questions

1 Copy and complete these sentences by choosing from the words in bold.

The Earth's axis is slightly **tilted/bent**. This tilt of the Earth causes the **tides/seasons** and day length.

In the UK it is summer when the northern hemisphere is tilted towards the **Sun/Moon**.

We have the **shortest/longest** day on 21 June and the **shortest/longest** day on 21 December.

2 Is the Earth tilted towards the Sun or away from the Sun on the shortest day?

3 Make a list of all the changes between winter and summer that we notice in the UK.

For your notes:

- It takes the Earth 365¼ days or a **year** to orbit the Sun.

- The Earth is slightly tilted on its axis. This gives us the **seasons**.

L4 Round the Sun

Orbiting the Sun

Our **Solar System** is made up of the Sun along with nine planets, including the Earth, all orbiting the Sun. Some of the planets have moons that orbit them, like the Moon orbits the Earth. There are also small lumps of rock called **asteroids** between Mars and Jupiter.

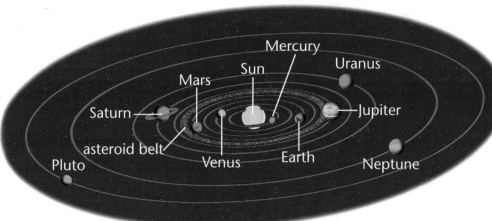

Mercury, Uranus, Sun, Mars, Saturn, Jupiter, asteroid belt, Venus, Earth, Neptune, Pluto

a Which planet is closest to the Sun?

b Which planet is furthest from the Sun?

c What are asteroids made of?

Planetary facts

The order of the planets

- Sun
- Mercury
- Venus
- Earth
- Mars
- Jupiter
- Saturn
- Uranus
- Neptune
- Pluto

Mercury is a small rocky planet that is closest to the Sun. It has no atmosphere and is very hot.

Earth is a warm, rocky planet with water and an atmosphere of nitrogen and oxygen. It is the only planet we know about with living organisms.

Venus is the hottest planet. It is rocky with an atmosphere of carbon dioxide.

Mars is a red, rocky, cold planet with an atmosphere of carbon dioxide.

Jupiter is a cold giant planet made mainly of liquids and gases. Jupiter is famous for its Great Red Spot. It has many moons.

Saturn is a cold giant planet made mainly from gases. It has beautiful rings around it and many moons.

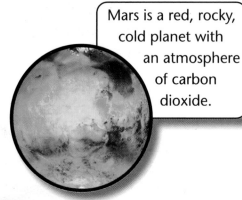

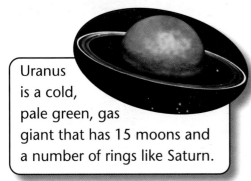

Uranus is a cold, pale green, gas giant that has 15 moons and a number of rings like Saturn.

Neptune is a bluish gas giant with a cold atmosphere.

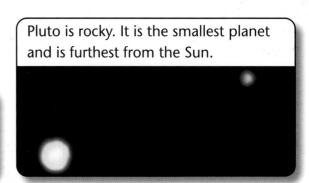

Pluto is rocky. It is the smallest planet and is furthest from the Sun.

d Read the planetary facts. Suggest a reason why it seems the Earth is the only planet suitable for life.

Planet years

The Earth takes $365\frac{1}{4}$ days, or a year, to orbit the Sun. Mercury is the closest planet to the Sun and has the shortest year of only 88 Earth days. Pluto is the furthest away and takes 248 Earth years to orbit the Sun. The length of a planet's year depends on how far it is from the Sun.

Collecting evidence

Astronomers have found out a lot about the Solar System using telescopes. In 1609 Galileo made the first telescope. This allowed him to see things in space magnified 30 times.

Later, when photography was invented, astronomers could take photographs through their telescopes. Pluto was discovered from a photograph in 1930. Modern telescopes like the Hubble telescope detect radio waves from distant stars and make pictures. These stars are too far away to send out visible light so we cannot 'see' them.

e How was Pluto discovered?

Do you remember?

The Sun is much bigger than the Earth and the Moon is smaller than the Earth. It is only in the last 40 years that we have had photographs from space to show this.

Questions

1 Make up a sentence to help you remember the order of planets in the Solar System. Use the first letter of each planet to start each word in your sentence.

2 Choose the correct planet for each fact and write them down.

Planets

| Earth | Jupiter | Mercury | Neptune | Venus |
| Saturn | Pluto | Uranus | Mars | |

Facts

Planet closest to the Sun   Only planet we know that has life

Planet furthest from the Sun   Planet with a Great Red Spot

Hottest planet

For your notes:

- The **Solar System** is made up of the Sun and the nine planets, along with their moons and **asteroids**.

- The planets differ from each other in many ways, such as distance from the Sun and what they are made of.

131

L5 Making models

The Solar System

Class 7J were studying the Solar System. They made models to show the Solar System. They used the information in this table. The Sun is 1 392 000 km in diameter.

Planet	Diameter (to nearest 1000 km)
Mercury	5000
Venus	12 000
Earth	13 000
Mars	7000
Jupiter	143 000
Saturn	121 000
Uranus	51 000
Neptune	50 000
Pluto	2000

Relative sizes

Ian, Karl and Darren made a model showing the sizes of the planets. Darren brought in a yellow beach ball to be the Sun. Karl and Ian brought in other balls to be the planets.

Ball	Diameter in cm
beach	40
basketball	23.9
football	22.3
netball	21.3
volleyball	20.7
cricket	7.3
tennis	6.4
squash	4.4
golf	4.3
table tennis	3.8

a Karl also brought in a rugby ball, but they did not use it. Why not?

They put the balls in order with the biggest first and the smallest last, as shown in the table. The beach ball stands for the Sun.

b Put the planets in order of size too. Then decide which ball should stand for each planet.

Ian, Karl and Darren showed their model to the rest of the class. The model shows the sizes of the Sun and the planets. The class **evaluated** the model. First, they said the good things about the model.

> The balls are spheres, like the Sun and the planets.

> Uranus and Neptune are almost the same size, and the netball and the volleyball are almost the same size.

c Suggest two other good points about the model.

The class then thought of ways to improve the model.

d Suggest two ways to improve the model.

Scale model

Ian, Karl and Darren decided to make a **scale** model. They made a ball of clay 2 mm in diameter for Pluto. They used 1 mm to represent 1000 km. So, the real diameter of Pluto at 2000 km is represented by 2 mm. They divided the real diameter by 1000.

Using the same scale, the diameter of Jupiter will be 143 mm and Earth will be 13 mm.

e What will be the diameter of:
  (i)   Saturn?
  (ii) Mars?
  (iii) the Sun?

f Suggest why this scale model of the planets is better than the first model they made using balls.

Pluto

2 mm diameter
represents
2000 km diameter

Earth

13 mm diameter
represents
13 000 km diameter

Questions

1 Study the information opposite about the Earth and the Moon.

The diameter of the Earth is 13 000 km.
The diameter of the Moon is 3500 km.

Draw a scale diagram, or make a model, of the Earth and Moon with a scale of 1 mm = 1000 km.

Glossary

A The short way of writing amps.

acid A solution that has a pH lower than 7.

adaptations Having features that help a living thing to survive in a particular place.

adapted A well-adapted organism has features that help it to survive in a particular place.

adolescence The time in a young person's life when physical and emotional changes happen.

afterbirth The placenta comes out of the uterus after the baby is born. It is called the afterbirth.

air resistance The friction a moving object makes with air.

alkali A base that dissolves in water, forming a solution with a pH greater than 7.

alternative energy resources Energy resources that are not fossil fuels.

ammeter A device that measures the current in an electrical circuit.

amphibians One of the groups of vertebrate animals. Amphibians lay eggs in water but breathe air. They have a smooth, moist skin.

amps Current is measured in amps.

animal cells The building blocks that make up all animals. Animal cells have a cell membrane, cytoplasm and a nucleus.

animals Living things that feed on other living things and move around.

anther The part of the stamen in a flower that makes the pollen.

arthropods Group of invertebrate animals with segmented bodies and jointed legs.

asteroids Rocky objects in space. In our Solar System most of the asteroids are found in the asteroid belt.

atom The smallest part of matter.

axis An imaginary line through the Earth from the North Pole to the South Pole.

balanced forces Two forces of the same size pulling in opposite directions.

base A substance that reacts with an acid and neutralises it.

battery Stores chemical energy. More than one cell connected together in an electrical circuit.

biomass The total mass of a living thing, not including the water. Biomass can be used as an energy resource.

birds One of the groups of vertebrate animals. Birds lay eggs with hard shells, look after their young and have feathers and wings.

braking distance The distance a car travels after the driver puts the brakes on but before it stops completely.

camouflage Features that help a living thing to blend in with its surroundings.

carbonates Substances that react with acid to produce carbon dioxide. Many rocks are made of carbonates.

carbon dioxide A gas that is produced when carbon burns and joins with oxygen, or when a carbonate reacts with an acid. Carbon dioxide turns limewater milky.

carnivore An animal that feeds on other animals.

carpel The female organ in a flower, that produces the egg cells.

cell (electrical) An object that changes chemical energy into electrical energy.

cells (in living things) Tiny building blocks that make up all living things.

cell division A cell splits into two to make more cells.

cell membrane A thin layer that surrounds the cell and controls the movement of substances in and out of the cell.

cell wall A tough box-like wall around plant cells.

cellulose A tough stringy substance found in plant cell walls.

centipedes One of the groups of arthropods. Centipedes have lots of legs and a segmented body.

cervix A ring of muscle at the opening of the uterus.

charcoal A fuel made from wood, that is mainly carbon.

chemical change A change that makes a new substance. Many chemical changes are irreversible.

chemical energy Energy stored in a material, which will be given out in a chemical reaction.

chemical reaction A change that makes a new substance.

chlorophyll A green substance that is needed for photosynthesis.

chloroplasts The parts of a plant cell that carry out photosynthesis.

chromatography A method used to separate mixtures of substances. The most soluble substances travel the furthest up the paper.

classification Putting things with similar features into the same group.

coal Material made from plants that lived many millions of years ago, used as a fuel.

combustion The chemical reaction that happens when something burns.

compensate/compensation Make up for a change by balancing things out.

complete circuit Cells and lamps or other things joined by wires to make a closed loop.

concentrated A solution that contains a lot of dissolved solute is concentrated.

condenses Changes from a gas to a liquid.

condensation Changing from a gas to a liquid.

conditions Something in a habitat that varies and can be measured, such as temperature or rainfall.

conserve/conserving Using fuels or other resources carefully, so they last longer.

conserved When the same amount is there at the end as there was at the beginning. For example, when a solid is dissolved in a liquid, the mass of solution equals the mass of solid plus the mass of liquid. The mass is conserved.

consumer An animal, that eats (consumes) plants or other animals.

contractions The muscles of the uterus wall squeeze when a baby is born.

cord This links the developing fetus to the placenta in a pregnant female animal.

corrosion Eating away of the surface of a solid by a chemical reaction.

corrosive Substances that may destroy living tissues on contact are corrosive.

crustaceans One of the groups of arthropods. Crustaceans have lots of legs, a soft body and usually a hard shell.

current Electricity flowing around the circuit.

cytoplasm A jelly-like substance found inside cells.

dense A dense material has a lot of particles in a small volume.

density How heavy a material is for its size.

diffuses A gas or liquid spreads out and mixes with the gas or liquid around it.

diffusion Gas or liquid particles spreading out as their particles move and mix.

dilute A solution that does not contain much dissolved solute is dilute.

displaces Pushes out or replaces.

distillation A method used to separate the solvent from a solution, or to separate mixtures of liquids with different boiling points.

distilled water Water that has been made pure. It has been changed to a gas and condensed back to a liquid again.

dormant An inactive state that allows an organism to survive harsh conditions, such as the winter.

dyes Coloured substances.

egg *or* **egg cell** The female sex cell in an animal or plant. The egg joins with the male sex cell in reproduction.

electrical energy Energy carried by electricity.

embryo A tiny ball of cells formed from the fertilised egg in animal reproduction.

energy Energy makes things work. When anything happens, energy is transferred.

energy transfer The movement of energy from one place to another.

energy transfer diagram A diagram with arrows that shows how energy moves from place to place.

environment The surroundings.

environmental variation Differences in features that are affected by our surroundings are examples of environmental variation.

evaluate To judge how good a model or experiment is, finding its good points and bad points.

evaporates Changes from a liquid to a gas.

expand To get bigger. A solid expands when you heat it because the particles move faster and take up more space.

extension The amount a spring stretches when you hang a weight on it.

fault Something that stops an electrical circuit from working.

features Special parts of organisms, or particular things they do.

fertilisation In an animal, a sperm joining with an egg to make a baby. In a plant, a pollen grain joining with an egg cell to make an embryo plant.

fetus A developing baby inside the uterus of a female mammal.

fire triangle A way of showing the three things a fire needs to burn – fuel, oxygen and energy.

fish One of the groups of vertebrate animals. Fish live in water and lay eggs there. They breathe through gills and have scales and fins.

flatworms One of the groups of invertebrate animals. Flatworms have a flat leaf-shaped body.

float An object floats when the upthrust is equal to its weight. It stays on the top of the water.

food chain A diagram that shows how the organisms in a habitat feed on each other.

food web Two or more food chains link together to form a food web, that shows the feeding relationships between the organisms.

force arrows Arrows we draw that point in the direction of a force. The length shows the size of the force.

fossil The remains of an animal or plant that has been buried deep underground for millions of years and preserved.

fossil fuels Materials made from animals and plants that lived many millions of years ago, used as a fuel.

friction The force that is made when things rub together.

fuel A material that has a lot of stored chemical energy. We burn a fuel to use the energy.

fungi Living things that feed on rotting material, for example, toadstools.

fuse A safety device for electrical circuits. The fuse has a very thin wire which melts if the current gets too high, and breaks the circuit.

galaxy A collection of millions of stars held together by gravitational pull.

gas A state of matter that is not very dense. A gas is easily squashed. Its shape and volume can change.

gas pressure A force caused by gas particles hitting the sides of their container.

gestation period The time a baby takes to develop inside its mother before it is born.

glands Parts that make hormones and other substances in animals. In male animals, the glands in the reproductive system make a liquid which mixes with sperm to make semen.

gravitational attraction (gravity) The force that pulls everything towards the centre of the Earth. The other planets, the Moon and the Sun also pull things because of gravitational attraction.

gravitational energy Energy stored because something is lifted up.

greenhouse effect The carbon dioxide in the air stops some of the heat energy escaping from Earth so making Earth warmer. It behaves like glass in a greenhouse.

growth Increase in size. Organisms grow by increasing the number of cells and increasing the size of the cells.

habitat The place where a living thing lives.

harmful Harmful substances may have a health risk similar to but less serious than toxic substances.

heat energy Energy transferred from a hot object to a cooler object.

herbivore An animal that feeds on plants.

hibernation An animal goes into a deep sleep to survive difficult conditions in the winter.

hormone A substance in the body that makes changes happen.

hydrogen An explosive gas.

identical twins Two babies that came from the same sperm and egg. They are born at the same time and they look exactly the same.

indicator A coloured substance that shows whether the solution being tested is acidic, alkaline or neutral.

inherited Passed on from the parents to their offspring.

inherited variation Differences in features that are passed on from the parents are examples of inherited variation.

input variable The thing you change in an investigation.

insects One of the groups of arthropods. Insects have six legs and a three-part body.

insoluble A substance that is insoluble will not dissolve.

interdependence Organisms in the same food chain all depend on each other.

invertebrates Animals without backbones.

irreversible change Something that cannot be changed back to how it was before.

irritant Substances that can cause redness or blistering if in contact with the skin are irritant.

J The short way of writing joules.

jellyfish One of the groups of invertebrate animals. Jellyfish have a soft jelly-like body.

joules Energy is measured in joules.

kg The short way of writing kilograms.

kilograms Mass is measured in kilograms.

kilojoules There are 1000 joules in 1 kilojoule.

kilometres per hour Speed may be measured in kilometres per hour.

kinetic energy The scientific name for movement energy.

kJ The short way of writing kilojoules.

km/h The short way of writing kilometres per hour.

leap year A year that has 366 days, that occurs every four years.

light energy Energy transferred by light.

light source Something that gives out light energy.

lime A basic substance containing calcium oxide, or other calcium compounds.

limewater A solution used to test for carbon dioxide. Limewater turns milky when carbon dioxide bubbles through it.

liquid A state of matter that flows. The shape of a liquid can change, but its volume is fixed.

litmus An indicator made from lichens. Acids turn blue litmus red. Alkalis turn red litmus blue.

lubricant A substance that reduces friction by making surfaces run smoothly against each other.

luminous Objects that give out light are luminous.

lunar eclipse An eclipse that occurs when the shadow of the Earth moves across the Moon.

m/s The short way of writing metres per second.

mammals One of the groups of vertebrate animals. Mammals have hairy skin. Their babies develop inside the mother and are fed on milk.

mammary glands Features that female mammals have, that make milk.

mass A measure of how much matter an object has.

material Anything that is made up of particles. A material may be a solid, a liquid or a gas.

matter Anything that has mass is made up of matter. Matter contains particles.

menstrual cycle A monthly cycle in women. During the cycle an egg is released, and the woman has a period.

methane A hydrocarbon fuel that is a gas. Natural gas is mainly methane.

metres per second Speed may be measured in metres per second.

microorganism A very small living thing that can only be seen with a microscope.

microscope A device that is used for looking at very small objects.

migration Moving to another habitat to avoid difficult conditions, for example, swallows fly south to avoid the cold winter in the UK.

millipedes One of the groups of arthropods. Millipedes have lots of legs and a segmented body.

model An idea or picture made up by a scientist to show a situation that cannot be seen. A model helps scientists think through explanations.

molecule A group of two or more atoms joined together.

molluscs One of the groups of invertebrate animals. Molluscs have a soft muscular body with one foot, and usually a hard shell.

movement energy When something moves, it has movement (kinetic) energy.

N The short way of writing newtons.

natural gas A gas formed from animals and plants that lived many millions of years ago, used as a fuel. It is mostly methane.

neutral A substance that is neither acidic nor alkaline, with a pH of 7, is neutral.

neutralisation The chemical reaction that takes place when an acid reacts with a base.

newtons Force is measured in newtons.

non-identical twins Two babies that came from different sperm and eggs. They are born at the same time, but look different.

non-luminous An object that does not give out light is non-luminous.

non-renewable energy resource An energy resource that is not replaced as we use it is non-renewable.

northern hemisphere The top half of the Earth, above the Equator.

nucleus The part of a cell that controls everything the cell does.

oil A liquid formed from animals and plants that lived many millions of years ago, used as a fuel.

omnivore An animal, that feeds on both plants and animals.

orbit The path a body takes around the object it is travelling round, such as the Moon's orbit around the Earth.

organ A group of different tissues that work together to do a job.

organism A living thing, that carries out the processes of life.

outcome variable The thing that changes during an investigation. The outcome variable is the thing you measure.

ovary In an animal, part of the female reproductive system that makes the eggs. In a plant, part of the carpel that makes the egg cells.

oviduct A tube in the reproductive system of a female animal. The eggs travel down the oviduct to the uterus.

ovulation An egg is released into the oviduct from the ovary.

oxide An oxide is made when a substance burns and joins with oxygen from the air.

oxygen A gas. Oxygen is used in burning and in respiration.

parallel circuit A circuit with more than one loop.

particle model The idea that everything is made up of particles.

particles Tiny parts that make up every type of matter.

penis Part of the reproductive system in a male animal. The penis allows the sperm to be placed inside the vagina.

period Part of a woman's menstrual cycle. The lining of the uterus breaks down and leaves the body through the vagina.

petal The part of a flower that is often colourful and attracts insects.

pH scale A number scale used to measure the strength of acidity and alkalinity.

phases of the Moon The different shapes of the Moon we see as the Moon orbits the Earth.

physical change A change in which no new substance is made. A change of state is a physical change. Physical changes are reversible.

placenta Structure formed in a pregnant female mammal. The developing fetus gets its food and oxygen from the placenta.

plant cells The building blocks that make up all plants. Plant cells have a cell membrane, cytoplasm and a nucleus, and also a cell wall, chloroplasts and a vacuole.

plants Living things that are green and make their own food using sunlight.

pollen grains The male sex cells in a plant. A pollen grain joins with an egg cell to make the seed.

pollen tube A tube that grows from the pollen grain on the stigma, down the style to the ovule, so that the pollen grain nucleus can reach the egg cell.

pollination The transfer of pollen from an anther to a stigma in plant reproduction.

predator An animal that hunts and feeds on other animals.

pregnancy The time when a female animal has a baby growing inside her uterus.

pregnant A female animal is pregnant when there is a baby growing inside her uterus.

prey Animals that are hunted and eaten by predators.

producer A plant, that produces its own food by photosynthesis.

products The new substances that are formed in chemical reactions.

puberty The first part of adolescence, when physical changes happen

pure A pure material only contains one substance.

reactants The substances that take part in a chemical reaction, and change into the products.

reaction force A force that stops things falling through solid objects. When you sit on a chair, your weight is balanced by the reaction force from the chair.

reflected When light bounces off a surface, it is reflected.

relationship A pattern that links variables together. A relationship describes how the outcome variable changes when the input variable is changed.

renewable energy resource An energy resource that can be replaced as we use it.

reproduce/reproduction To make more organisms of the same species.

reptiles One of the groups of vertebrate animals. Reptiles breathe air and lay eggs on land. They have a scaly, dry skin.

resistance How much something slows down the electric current passing through it. A thin wire slows down the current more than a thick wire, so it has a higher resistance.

reversible change Something that can be changed back to how it was before.

ring main The electrical circuit in a house. It is a parallel circuit.

roundworms One of the groups of invertebrate animals. Roundworms have a soft, thin round body.

sample A small part of something, used to represent the whole.

saturated A solution is saturated when no more of a solute can dissolve in it.

scale A scale drawing or model shows something bigger or smaller than it really is.

scale diagram A drawing that shows something bigger or smaller than it really is.

scale factor A number used in scale drawing. You multiply by the scale factor to scale something up. You divide by the scale factor to scale it down.

scaling down Making something smaller.

scaling up Making something bigger.

scrotum Part of the reproductive system in a male animal. The scrotum is a bag of skin that holds the testes.

seasons Times of different climate during the year. In the UK we have four seasons – spring, summer, autumn and winter.

seed A structure made in a flower, that contains the new plant and a food store.

segmented worms One of the groups of invertebrate animals. Segmented worms have a soft ringed body.

segments Sections of the body in arthropods and segmented worms.

semen A mixture of sperm and a special liquid to help them swim.

series circuit A circuit in which everything is in one loop.

sexual intercourse The man's penis enters the woman's vagina, and sperm are released into the vagina.

solar eclipse An eclipse that occurs when the Moon blocks the Sun's light from reaching the Earth. A shadow passes across the Earth.

solar energy Energy given out by the Sun.

solar furnace A device that concentrates heat (thermal) energy from the Sun and uses the heat energy to heat a material or to generate electricity.

Solar System The Sun and the objects orbiting it, including Earth and the other planets.

solid A state of matter that is dense and has a fixed shape and volume.

solubility A measure of how much of a solute will dissolve at a particular temperature.

soluble A substance that is soluble will dissolve.

solute The substance that dissolves to make a solution.

solution A mixture of a solute dissolved in a solvent.

solvent A liquid that substances can dissolve in.

sound energy Energy transferred by sound.

species A particular type of animal or plant. Members of a species can reproduce to form more of their kind.

speed How fast something is moving.

sperm The male sex cells in an animal. The sperm joins with the egg in reproduction.

sperm tube A tube in the reproductive system of a male animal. Sperm swim from the testis to the penis through the sperm tube.

spiders One of the groups of arthropods. Spiders have eight legs and two parts to the body.

stamens The male organs in a flower, that produce the pollen.

starfish One of the groups of invertebrate animals. Starfish have a hard star-shaped body.

stigma The part of the carpel where the pollen grain lands.

stopping distance The distance a car travels after the driver decides to stop but before it stops completely. Stopping distance = thinking distance + braking distance.

strain energy Energy stored in a material because the material is being pulled or pushed.

style The part of the carpel that holds up the stigma.

testis Part of the reproductive system in a male animal. The **testes** (plural) make the sperm.

theory A set of ideas to explain something.

thermal energy The scientific name for heat energy.

thinking distance The distance a car travels after the driver decides to stop but before he or she puts the brakes on.

tissue A group of similar cells that carry out the same job.

toxic Toxic means poisonous. Substances that may cause serious health risks and even death if inhaled, taken internally or absorbed through the skin are toxic.

transferred Moved from one place to another.

twins Two babies that develop together inside the mother and are born at the same time.

unbalanced forces Forces pushing in different directions when one force is bigger than the other. An unbalanced force makes the object move or speed up or slow down.

universal indicator An indicator that has a range of colours showing the strength of acidity or alkalinity on the pH scale.

upthrust The force caused by water pushing up against an object.

uterus Part of the reproductive system in a female animal. The baby grows and develops in the uterus.

V The short way of writing volts.

vacuole A bag inside plant cells that contains a liquid which keeps the cell firm.

vagina Opening to the reproductive system in a female animal. Sperm enter the woman's body through the vagina, and the baby leaves through the vagina when it is born.

variable A thing that we change or that changes in an investigation.

variation The differences between living things, or between members of a species.

vertebrates Animals with backbones.

voltage How much energy the electricity is carrying.

volts The energy a cell stores is measured in volts.

water turbine A device that takes in the movement (kinetic) energy of falling water and gives out electrical energy.

water vapour Water that has turned to a gas.

wave energy The movement (kinetic) energy of waves.

wave turbine A device that takes in the movement (kinetic) energy of waves and gives out electrical energy.

weight The force of gravitational attraction on an object, that makes it feel heavy.

wind energy The movement (kinetic) energy of the wind.

wind turbine A device that takes in the movement (kinetic) energy of the wind and gives out electrical energy.

word equation An equation in words to show a chemical reaction.

year The time taken for the Earth to orbit the Sun.

Index

Note: page numbers in **bold** are for glossary definitions